CAMBRIDGE LIBRARY COLLECTION

Books of enduring scholarly value

Perspectives from the Royal Asiatic Society

A long-standing European fascination with Asia, from the Middle East to China and Japan, came more sharply into focus during the early modern period, as voyages of exploration gave rise to commercial enterprises such as the East India companies, and their attendant colonial activities. This series is a collaborative venture between the Cambridge Library Collection and the Royal Asiatic Society of Great Britain and Ireland, founded in 1823. The series reissues works from the Royal Asiatic Society's extensive library of rare books and sponsored publications that shed light on eighteenth- and nineteenth-century European responses to the cultures of the Middle East and Asia. The selection covers Asian languages, literature, religions, philosophy, historiography, law, mathematics and science, as studied and translated by Europeans and presented for Western readers.

A Synopsis of Science

James Robert Ballantyne (1813–64) taught oriental languages in India for sixteen years, compiling grammars of Hindi, Sanskrit and Persian, along with translations of Hindu philosophy. In 1859, for the use of Christian missionaries, he prepared a guide to Hinduism, in English and Sanskrit (also reissued in this series). Published in two volumes in 1852, *Synopsis of Science* was intended to introduce his Indian pupils to Western science by using the framework of Hindu Nyaya philosophy, which was familiar to them and which Ballantyne greatly respected. This second volume proceeds in similar style to the first: through a series of short paragraphs, Ballantyne introduces arithmetic, algebra, calculus, chemistry, botany, zoology, anatomy, physiology, mineralogy and geology. The second part of the volume is a Sanskrit translation. Overall, the work serves as an excellent primary source on the educational aspects of British imperialism.

Cambridge University Press has long been a pioneer in the reissuing of out-of-print titles from its own backlist, producing digital reprints of books that are still sought after by scholars and students but could not be reprinted economically using traditional technology. The Cambridge Library Collection extends this activity to a wider range of books which are still of importance to researchers and professionals, either for the source material they contain, or as landmarks in the history of their academic discipline.

Drawing from the world-renowned collections in the Cambridge University Library and other partner libraries, and guided by the advice of experts in each subject area, Cambridge University Press is using state-of-the-art scanning machines in its own Printing House to capture the content of each book selected for inclusion. The files are processed to give a consistently clear, crisp image, and the books finished to the high quality standard for which the Press is recognised around the world. The latest print-on-demand technology ensures that the books will remain available indefinitely, and that orders for single or multiple copies can quickly be supplied.

The Cambridge Library Collection brings back to life books of enduring scholarly value (including out-of-copyright works originally issued by other publishers) across a wide range of disciplines in the humanities and social sciences and in science and technology.

A Synopsis of Science

From the Standpoint of the Nyaya Philosophy

VOLUME 2

JAMES R. BALLANTYNE

CAMBRIDGE UNIVERSITY PRESS

Cambridge, New York, Melbourne, Madrid, Cape Town,
Singapore, São Paolo, Delhi, Mexico City

Published in the United States of America by Cambridge University Press, New York

www.cambridge.org
Information on this title: www.cambridge.org/9781108056335

© in this compilation Cambridge University Press 2013

This edition first published 1852
This digitally printed version 2013

ISBN 978-1-108-05633-5 Paperback

A
SYNOPSIS OF SCIENCE;

FROM

THE STANDPOINT

OF

THE NYÁYA PHILOSOPHY.

SANSKRIT AND ENGLISH.

VOL. II.

PRINTED FOR THE USE OF THE BENARES COLLEGE
BY ORDER OF GOVT. N. W. P.

————

Mirzapore:

ORPHAN SCHOOL PRESS :—R. C. MATHER, SUPT.

1852.

[1st Edition ;—500 copies,—Price 14 as.]

PREFACE.

In the Third Book of the Synopsis an attempt to establish a purely Hindú nomenclature for the science of Chemistry has been commenced. Mr. Mack, in the preface to his very careful treatise on Chemistry published in Bengálí and English at Serampore, in 1834, tells us that he was advised to discard all European terms in his Bengálí version, but that he could not persuade himself to adopt the advice. He retained therefore many of the European names, and adapted Sanskrit terminations to them. Several of these terminations have been adopted in the present essay, but the European names have been entirely discarded. As an educational instrument,—and it is in *this* capacity that we at present seek to employ it,—the science of Chemistry loses more than half its value when its compound terms do not tell their own meaning; and it is impossible that they should rightly tell their own meaning to one who is not familiar with the language from which they are derived. To an Englishman, unacquainted with the classical languages, the study of a work on Chemistry is very far from being such a mental exercise as it is to a classical scholar. The long compound names which exercise the reflection and excite the admiration or provoke the criticism of the latter, more frequently torture the memory and bewilder the understanding of the former. How entirely is the scientific beauty of the nomenclature thrown away upon the man who must look out Hydro-chlorate and Sesquioxide in his glossary in order to make sure which is which. It is all very well to teach long chemical names by rote to a youth who is to be employed as an apprentice in wielding a pestle. Him you perhaps do not seek to educate ; you merely make a convenience of him ; and if he does not practically mistake Corrosive Sublimate for Coloquintida in making up a prescription, why all is well. The case is otherwise where the aim is to educate and to instruct. Where Chemistry is to be efficiently employed for such a purpose, the learner must be conversant with Latin and Greek, or else the language of the science must be rendered into the language of the learner, as has been in a great measure done by the Germans for themselves.

The first question, in settling a chemical nomenclature, regards the naming of the Simple Bodies. The common metals, as well as sulphur and carbon, have names in most languages which there is no occasion for changing. All the others Simple Bodies require to have names devised for them. First there are the four simple gases. The name of Oxygen—' the generator of acids'— might readily be rendered by a corresponding Sanskrit compound ; but this (as Mr. Mack has remarked) would only tend to preserve the exploded theory that there is no generator of acids besides Oxygen. Its old name of Vital Air connotes one of its most important characters, and therefore it has been here named *práṇa-prada*, or *pránaprada-váyu*,—' the air that emphatically gives us breath.' Nitrogen (or Azote) we call *jivántaka*, 'that which would put an end to life.' Hydrogen is *jalajanaka*, ' the water-former ; and Chlorine *harita* ' the greenish-coloured.'

Of the nine simple non-metallic bodies that are not gaseous, two, viz. Sulphur (*gandhaka*) and Carbon (*angára*), have Sanskrit names. Boron, as it is the basis of borax (*ṭanka*), we therefore call *ṭankopádána* ; Silicon is the generator of flint,—*agnipras-thara-janaka* ; Selenium,—so named after the moon,—we have likewise named after the moon,—*chándra*,—it being a matter of moonshine what so rare and unimportant a substance be denominated. Phosphorus is *prakásada* 'the giver of light;' Bromine is *púta* 'the fetid;' Iodine is *aruṇa*,—the name, like the Greek one, referring to the violet-colour of its vapour; and Fluorine is *káchaghna-janaka* ' the generator of that (fluoric acid) which corrodes glass.'

Of those metals which have no names in Sanskrit only six are named in the present essay. Platinum, the 'heaviest' of metals, is, with allusion to its weightiness, named *gurutama*; and Potassium, the ' lightest,' *laghutama*. Sodium is 'the basis of culinary salt'—*lavaṇa-kara*; and Calcium, ' the basis of nodular limestone,' —*śarkará-kara*. Zinc, the Urdú name of which is *dastá*, we have named *dasta*—with allusion to the way in which its oxide, the ' philosophical wool ', is ' tossed about' in the air. Tin, from its resemblance to silver in colour, we call *rajata-práya*.

Let such, then, be the names that we have to deal with in forming the names of Compounds ;—and first of Binary compounds. Compounds must have names suggestive of the fact that they are *acid* or otherwise. The termination *ic* belongs to the Sankrit as well as to the Latin,—so that Sulphur and Sulphuric Acid can be satisfactorily rendered *gandhaka* and *gándhakik-ámla*. To the acids in *ous* another termination (*ya*) has been appropriated. To the non-acid binary compounds, without attempting at present to fix separate terminations for the several varieties, the general termination *ja*—meaning 'produced from'—has been assigned. Thus an Oxide is *práṇaprada-ja*; a Chloride *harita-ja*; and so on. The Alkalis—potassa and soda—take feminine names, according to the analogy of the Latin, from those of their metallic bases,—thus—*laghutamá* and *lavaṇa-kará*.

Coming to the compounds of compounds,—as the acid affix *ic* changes to *ate* in the name of the resulting salt, the Sanskrit *ika* is replaced by *áyita*. Thus, as the Sulphuric Acid gives a Sulphate, the *gándhakik-ámla* gives a *gandhakáyita*. It should be unnecessary to remark, that the suitableness of these names is not to be estimated on the principle which led the British sailor to set down the Spaniards as a nation of fools because they call a hat a *sombrero*. To the British sailor the word hat sounds much more natural than sombrero ; and for like reasons Sulphate of Soda may seem to sound much more natural than *lavaṇa-karáyá gandhakáyitam*. But as ' hat' is not good Spanish, so ' Sulphate of Soda' is not good Sanskrit ; and this leads us to forestall another criticism of kindred calibre. Is the sombrero-like expression, *lavana-karáyá gandhakáyitam*, good Sanskrit? The question is not to be resolved by submitting the term to a Sanskrit grammarian ignorant of physical science,—to whom, without an attentive, serious, ingenuous, and uncavilling study of the tract in which it appears, the term has a *right* to be as obscure as the term Binoxalate of Potassa to the grandfathers of Lindley Murray.

It is intended that the present sketch of the science, after its phraseology shall have been more fully tested, shall form the text for a commentary in which the topics will be more fully treated, —directions being furnished also for the performance of experiments and the conducting of processes illustrative of the doctrine.

Benares College,   }   J. R. B.
1st July 1852.   }

BOOK THIRD.

SECTION I.

INTRODUCTORY.

a. At the end of the First Book we mentioned the reason for now proceeding to consider the properties of Number and Magnitude abstractedly considered,—that is to say, independently of any consideration of the particular things numbered or measured.

b. The science of determinate numbers is called Arithmetic.

SECTION II.

ARITHMETIC.

c. The following is probably what suggested the employment of the decimal notation.

Aphorism I.

Numbers are made to increase by tens because men have ten fingers.

a. On looking at an assemblage of objects,—say of men,—if the spectator wishes to ascertain or to remember the number, he may use his fingers as the means of notation. Let him hold up a finger for each man, and give a name, "one," "two," "three," and so on, to each finger. When he has come to the last finger, he must begin a new set, and he must note in some way how many sets there were in all. It is probable that this was the origin of the method of notation in general use, where the values of the figures, in the successive places, increase by tens. If men had had twelve fingers, the values of the figures would probably have been made to increase by twelves.

b. The rules for calculation derived from this science may be studied in Bápú Deva's treatise.

c. The ordinary rules of computation are inconvenient when we have to deal with very large numbers. In such cases we may make use of the method of *Logarithms*, some account of which here follows.

Aphorism II.

By means of Logarithms the processes of multiplication and division can be reduced to those of addition and subtraction; and involution and the extraction of roots can be effected by a single multiplication or division.

a. Take a set of numbers, each number exceeding the number preceding it by a given difference, and let the first number in the set be equal to the given difference; as 1, 2, 3, 4, 5, &c., where the difference is 1. Then take another set of numbers such that each is equal to twice or thrice or any fixed number of times the one before it, the first number in the set being equal to the multiplier, as 10, 100, 1000, 10000, &c., where the multiplier is 10. Write the second set of numbers under the first set, thus—

1	2	3	4	5	6	7
10,	100,	1000,	10000,	100000,	1000000,	10000000,

then the numbers in the upper line will be the Logarithms* of those in the lower, and the facts asserted in the Aphorism will be found true as follows.

b. Take the sum of two of the numbers in the upper line, and there will be found in the lower line opposite that number the product of the numbers which stand in the lower line opposite to the numbers the sum of which was taken. For example, of the two numbers 2 and 4 in the upper line, the sum is 6 ; and under 6 is found the number 1000000; and this is equal to the product of the numbers 100 and 10000, which stand beneath 2 and 4.

c. Again, if you take the difference of two numbers standing in the upper line, then you will find in the lower line, opposite that difference, the quotient which arises by dividing the greater by the lesser of the numbers standing under those the difference of which was taken. For example, the difference of 6 and 4 is 2, and under 2 you will find 100, which is the quotient of 1000000 divided by 1000, the two numbers standing under 6 and 4.

d. Again, if you multiply or divide, by any number, one of the numbers in the upper line, then you will find, under the number thus obtained, the corresponding power or root of the number standing under the number multiplied or divided. For example, multiply by 3 the number 2 in the upper line, and under the product, 6, you will find 1000000, which is the third power of 100, the number standing under 2.

e. The logarithms of all the numbers from 1 up to many thousands have been calculated and arranged in tables, from which

---

* The word Logarithm means " the number of the ratios." If the ratio is 10 then $10^1=10$, or the logarithm of 10 is 1; $10^2=100$, or the logarithm of 100 is 2, 1 being to 100 in the *duplicate* ratio of 1 to 10, and so on.

the logarithm of any particular number can be readily ascertained.

Aphorism III.

Symbols of processes are employed for shortness.

a. That is to say, certain signs are employed in order to make the language of Arithmetic shorter. A cross placed between two numbers signifies that they are to be added together. Thus 2+4 means that 4 is to be added to 2.

b. Two parallel lines signify equality. Thus 2+4=6 means that 4 added to 2 is equal to 6.

c. A single line between two numbers signifies that the latter is to be subtracted from the former. Thus 6—2=4, six diminished by 2, is equal to 4; and it is clear that 6, diminished by any number except 2, cannot be equal to 4.

d. It often however becomes necessary to speak of something which is true, not of any one number only, but of all numbers. For example, take 5 and 2; their sum is 7, their difference is 3. If this sum and difference be added together, we get 10, which is twice the greater of the two numbers first chosen. If from the sum we subtract the difference, we get 4, which is twice the lesser of the two numbers. The same thing will be found to be true of any two numbers.

e. Hence if we have two sealed bags of rupees, the one containing a large number, and the other containing a small number, both numbers being unknown to us, we can assert with certainty that the sum of the rupees in both bags, added to the difference between the two, is equal to twice the number of rupees contained in the larger bag; and that the sum, diminished by the difference, is equal to twice the number of rupees in the smaller bag.

f. Taking the letter *a* to represent the unknown number of rupees in the larger bag, and b to represent the unknown number in the smaller bag, we can assert that

$$(a+b)+(a-b)=2a, \text{ and } (a+b)-(a-b)=2b.$$

c. Such an employment of letters to denote quantities respecting which he can make assertions without knowing what the quantities are, has given rise to the science of Algebra.

SECTION III.

ALGEBRA.

Aphorism IV.

In Algebraical calculations the results are not particular but general.

a. In addition to the symbols already mentioned, the following are constantly employed in algebraical notation. A cross placed diagonally between two numbers signifies that they are to be multiplied together. Thus $a \times b$ means that the number denoted by a, whatever number it may be, is to be multiplied by the number which b stands for. This may be also represented by writing the letters together, thus ab. If $a=2$, and $b=3$, then $ab=6$.

b. When two letters are written, the one above a line and the other below it, it is signified that the upper is to be divided by the lower. Thus if $a=6$, and $b=2$, then $\dfrac{a}{b}=3$.

c. A number multiplied by itself, thus aa, is written thus a^2 This is called the square of a.

d. If again multiplied by itself, thus *aaa*, it is written thus a^3. This is called the cube of *a*.

e. The small figure written above the letter is called the index.

Aphorism V.

As the results of Algebraical operations are general, they furnish arithmetical rules.

a. For example : —when there are four numbers, the first of which contains the second exactly as many times as the third contains the fourth, the four numbers are said to be in proportion. For example, let the four numbers be represented by *a b c* and *d*; if *a* contains *b* exactly as many times as *c* contains *d*, then *a b c d* are in proportion. Suppose that *a* contains *b* three times, then $\dfrac{a}{b}=3$, and also $\dfrac{c}{d}=3$; and, as things that are equal to the same thing, are equal to one another, $\dfrac{a}{b}=\dfrac{c}{d}$.

b. Again, as the equimultiples of equals are equal, multiplying each by *bd*, we get $ad=bc$; that is to say, the product of the first and fourth, out of four numbers in proportion, is equal to the product of the second and third. This is the foundation of the very useful rule called the Rule of Three. In the Rule of Three, the first three terms of a proportion are given, and we are required to find the fourth. After determining the order of the terms from the conditions of the question, we multiply together the second and third. The product, we know, is the same as the product of the first and fourth, so that, in order to find the fourth, we have only to divide the product by the first term.

Aphorism VI.

The application of algebraical processes enables us to make discoveries respecting the properties of form such as would have been otherwise beyond our reach.

a. We shall briefly indicate the way in which this application is made. Suppose that we call an inch *one*—then a line of two inches will be represented by 2, a line of eight inches by 8, and a line of as many inches as are signified by the letter *a* will be represented by *a*. In like manner surfaces such as a parallelogram, and solids, such as a cube, can be represented by letters; and by performing algebraical operations on these symbols, new properties of the magnitudes are deduced.

b. There are some relations of form however which cannot be determined by the use of signs representing numbers. For example, the relation of the angles at the base of an isosceles triangle. Such relations of magnitude must be studied by the aid of lines, straight or curved; and this branch of the science of Quantity is entitled Geometry.

SECTION IV.

GEOMETRY.

c. Here a leading distinction among lines requires to be noticed at the outset.

Aphorism VII.

A straight line is one which never changes its direction. A curved line is one which continually changes its direction, so that no two successive points in it are in a straight line.

a. Hence if a point in a curve coincides with a point in a straight line, it is impossible that the next points should coincide.

b. In the Elements of Euclid, the text-book of elementary Geometry, which has been published in Sanskrit by Pandit Bápú Deva, the only curve line admitted is the circumference of the circle.

c. The postulate on which Euclid's reasonings are based is this, that two figures are equal, the one of which either exactly covers the other, or can be divided into portions which, all taken together, will exactly cover it.

d. Thus, in the fourth proposition of his first book, he infers that two triangles are equal which have two sides and the contained angle respectively equal, because the one triangle, if laid upon the other, must exactly cover it.

c. But here a difficulty arose. We have already explained that if a point in a curve coincide with a point in a straight line, the next points in the two lines cannot possibly coincide. Hence when we come to compare surfaces bounded by straight lines with surfaces bounded by curve lines, we may go on for ever subdividing the one or the other without being able to bring about a coincidence of their boundaries.

d. As it was of great importance to the advancement of Astronomy and other sciences dependent upon the aid of Mathematics, that we should be enabled to carry out the comparison of rectilinear with curvilinear surfaces and solids to any degree of nicety which might be required, a number of ingenious methods were invented during the two thousands years between the time of Euclid and the time of Newton.

e. That which has superseded all the rest, the method of Newton, improved by the notation of the German philosopher

SECTION V.

THE DIFFERENTIAL AND INTEGRAL CALCULUS.

f. In the *Lílávatí* of Bháskara A'chárya, the following question is proposed. A peacock seated on the top of a pillar sees at some distance a snake, which tenants a hole at the foot of the pillar. The snake, at the sight of its enemy, makes for its hole, and at the same instant the peacock flies down in pursuit of it. The height of the pillar being given, and the distance of the snake from the foot of it, and also the comparative speed of the two animals, it is required to determine the point at which the peacock will pounce upon the snake:—Now this question cannot be solved without the aid of the Calculus. Bháskara A'chárya, perceiving that there is a point at which the peacock will meet the snake if he fly down in a straight line, assumes that he will fly in a straight line, and proceeds to deduce the consequences of the supposition. But if the peacock be not such a mathematician as to know this point, he will direct his flight continually towards the snake, and as the snake is continually changing its place, the direction of the peacock's flight will continually change, and this, as previously stated, is the characteristic of a curved line. The determination of the curve in question involves processes intelligible only to those who have attained to some proficiency in the study of the Calculus.

g. The solution of the following simpler case may be rendered intelligible to one who understands at least the common rules of Algebra.

h. It has been ascertained by experiment that a body dropped from a height falls towards the earth through rather more than 16

B

feet in the first second, four times as much, or 64 feet, in two seconds, nine times as much, or 144 feet, in three seconds, and so
on—the space increasing according to the square of the time. A
very little consideration of this fact must make it apparent that
the rate of falling never continues uniform for even the smallest
assignable portion of time; else, why should the rate, after having been uniform for any period, become changed without the
intervention of a new cause? Now the rate of motion is always
estimated by the distance gone over by some assumed uniform
motion, in a given line ; and suppose we wish to know the rate
at which the body was falling exactly at the end of the third second, the rate must be something greater than it was immediately before, and something less than it was immediately
afterwards. We must here premise the first proposition of Newton's *Principia,* which is as follows.

Aphorism VIII.

" Quantities which tend towards equality, and of which the
difference, in the course of this approach, becomes less than any
finite magnitude that can be assigned, must be ultimately
equal."

a. " If this be denied, let them ultimately differ by a magnitude denoted by D. Then the difference between them cannot
become less than this finite magnitude D, which is contrary to
the hypothesis."

b. Now let us call 16 feet a *measure* ; then the number of measures through which the stone falls in x seconds is x^2. Let us
next suppose a very small portion of time h, and let the position
of the stone, which fell from the point O, be A at the end of x
seconds, B at the end of $x+h$ seconds, C at the end of $x+2h$ se-

conds, &c. Then the values of the lines, expressed in measures, are as follows :—

$$OA = x^2, \quad OB = (x+h)^2, \quad OC = (x+2h)^2, \quad \text{&c.} \quad |\text{—O}$$

Whence   —A

$$AB = OB - OA = 2xh + h^2 = (2x+h)\,h \qquad \text{—B}$$
$$BC = OC - OB = 2xh + 3h^2 = (2x+3h)\,h \qquad \text{—C}$$
$$CD = OD - OC = 2xh + 5h^2 = (2x+5h)\,h, \quad \text{&c.} \qquad \text{—D}$$

c. From this we see that in the successive small portions of time denoted by h, the spaces traversed were in the ratio of

$$2x+h, \quad 2x+3h, \quad 2x+5h, \quad \text{&c.}$$

d. Since, the greater the space fallen through, the greater is the change of the rate, it follows that the smaller the space fallen through, the smaller is the change of the rate ; or in other words the nearer does the rate approach to the condition of uniform motion.

e. The smaller then we suppose h to be, the more nearly will the foregoing expressions denote the uniform rate of motion at a given instant; and, according to the proposition quoted above, the expressions must ultimately coincide with the expression which we are in search of, if we carry the diminution of h to its extreme limit, or make it equal to *nothing.*

f. Now when h is made equal to nothing, all the above expressions become $2x$. So we find that the rate, at the end of any number of seconds, is twice that number of measures.

g. At the end therefore of three seconds when $x = 3$, the stone, had its motion suddenly become uniform, would have passed, in the course of the next second, through 96 feet; for $2 \times 3 \times 16 = 96$; whereas, in consequence of its continual acceleration, it actually passes through 112 feet.

h. The line OA, the length of which is here supposed to vary by successive additions, is called a variable quantity.

i. Variable quantities, in this branch of enquiry, are represented by the letters x y z.

j. The portions A B, B C, C D, by which the successive values of the variable differ from each other, and which may be supposed indefinitely small, are called differentials.

k. The differential of x is written thus dx; and, in like manner, those of y and z, dy and dz.

l. The question which has been worked out above, would be proposed in the Differential Calculus in the following shape. "If x vary uniformly, at what rate will x^2 vary?" We have seen that it will vary $2x$ times as fast; so that if the differential of x be dx that of x^2 is $2xdx$.

m. By comparing $2xdx$ with x^2 we arrive at the first rule for differentiation, viz. " Multiply by the index, diminish the index by unity, and multiply by the differential of the independent variable. "

n. The following are examples of differentiation;—$d. x^3$, that is to say the differential of the cube of x, is equal to $3x^2dx$; and again, $d.ax^4=4ax^3dx$.

o. The coefficient, or multiplier, of dx in these expressions is called the *differential coefficient.*—Thus, in $4ax^3dx$, the differential coefficient is $4ax^3$.

p. When the value of a quantity depends upon the particular value of another variable quantity, the one quantity is said to be a *function* of the other.

q. Thus, the area of a square, depending on the length of its side, is said to be a *function of the side.*

r. The object of the Differential Calculus is to determine the differential of any given function.

s. The object of the Integral Calculus is the converse of this, viz. to ascertain the function from which a given differential has been derived.

t. The first rule for integration is therefore the converse of that for differentiating.—viz." Add one to. the index, divide by the index thus increased, and by the differential of the variable."

u. By thus treating the differential $2xdx$ we recover the function x^3.

v. The symbol of integration is a long shaped *s,*—thus $\int 2xdx = x^2$.

SECTION VI.

PARENTHETICAL.

w. By means of the various methods of calculation which we have now briefly characterised, we are in many cases able to arrive deductively, or *à priori,* at a correct prediction of what will arise from the conjunct agency of several causes ; —e.g. of two forces acting on a single body.

x. In order that we may correctly predict the result of the conjunct agency of several causes, it is only necessary that the same law which expresses the effect of each cause acting by itself shall also correctly express the part due to that cause, of the effect which follows from the two together.*

* The paragraphs in this section are mostly taken from Mr. John S. Mill.

y. This condition is realized in the extensive class of phenomena called mechanical, namely the phenomena of the communication of motion (or of pressure, which is tendency to motion,) from one body to another.

z. In this class of cases of causation, one cause never, properly speaking, defeats or frustrates another. If a body is propelled in two directions by two forces, one tending to drive it to the north, and the other to the east, it is caused to move in a given time exactly as far in *both* directions as the two forces would separately have carried it; and is left precisely where it would have arrived, if it had been acted upon first by one of the two forces, and afterwards by the other.

aa. This law of nature is called the principle of the Composition of Forces; and the name of the Composition of Causes may be given to the principle which is exemplified in all cases in which the joint effect of several causes is identical with the sum of their separate effects.

ab. The principle of the Composition of Causes by no means prevails in all departments of the field of nature. The combination of two substances often produces a third substance with properties entirely different from those of either of the two substances. For example, when mercury, which is a fluid metal with a silvery colour, combines with sulphur, the colour of which is yellow, the substance produced, viz. vermilion, is of a bright red. Again, when lead, which has no taste, combines with vinegar, which has a very sour taste, a poisonous white salt is produced, which, from the sweetness of its taste, is vulgarly called sugar of lead.

ac. The investigation of the laws to which inanimate substances conform when their properties in combination are not the sum of their properties separately, is the province of Chemistry.

SECTION VII.

CHEMISTRY.

ad. In this science, bodies are regarded first under the following division.

Aphorism IX.

Bodies are Compound or Simple.

a. A Compound body is such as vermilion [§ VIII. *ab.*], which, for example, is compounded of mercury and sulphur.

b. That body, such as sulphur, which, so far as we can discover, is not compounded of any other bodies, we call a simple body.

c. Compound bodies are innumerable. These are all formed out of the simple bodies, of which there are reckoned above fifty five.

Aphorism X.

The Simple Bodies are divided into Metals and Non-metallic bodies.

a. The Metals are such as Gold, Iron, Mercury &c. The Non-metallic bodies are Sulphur &c.

Aphorism XI.

The Non-metallic simple bodies are divided into those that are in the form of Air, and those that are not so.

a. By an Air we mean a fluid indefinitely elastic, and permanently so at ordinary temperatures and under ordinary pressure.

The steam of boiling water is indefinitely elastic, but it loses its elasticity at ordinary temperatures, and this distinguishes it from Airs.

b. Airs are divided into the common Air which we breathe and other Airs,—these latter, for the sake of distinction, being called Gases.

Aphorism XII.

The simple Airs, or Gases are four in number.

a. These are named Oxygen (—also 'vital air'—), Nitrogen (—or Azote—which means incapable of sustaining life) Hydrogen (i. e. 'the former of water'), and Chlorine (i. e. the ' greenish yellow' gas).

b. Oxygen is called ' vital air' because men cannot live without it. Oxygen has no taste and no smell; and in a glass vessel full of it, we can discern no more colour than we can discern in a vessel when filled with common air. What, then, is the difference between common air and Oxygen ? One difference is the superior vivacity with which inflammable bodies burn in Oxygen.

c. A piece of wood with a spark of fire just on the point of being extinguished, will commence blazing if plunged into Oxygen.

d. Even iron burns in it like tinder.

e. Phosphorus, an inflammable substance obtained from bones, burns in Oxygen with a brilliancy like that of the sun.

f. We have said that men cannot live without Oxygen. But yet men live in common air. How is this ? Because there is

Oxygen in common air. But if there is Oxygen in common air, why does a piece of wood with a spark not begin blazing, and why does iron not blaze when heated ? It is because the Oxygen in the common air is mixed with and diluted by another gas,— viz. with Nitrogen.

g. Nitrogen, like Oxygen, has no taste, and no smell, nor can any colour be discerned in a glass vessel filled with it. What then is the difference between this gas and Oxygen ? One difference is this, that an inflamed substance is extinguished when it is placed in pure Nitrogen. If, therefore, the common air had consisted of Nitrogen alone, nothing could have burned ; and if it had consisted of Oxygen alone, every spark would have led to a conflagration. By the mixture of the two gases, in the proportions in which they exist in the atmosphere, both of these inconveniences are avoided.

h. The proportion of Nitrogen to Oxygen in the atmosphere is that of four to one. This may be illustrated as follows. Place a light, so that it may float, on the surface of a vessel of water. Invert over it a large bell-glass, such as a wall-shade, so that the edge of the glass may dip a little under the water. In a little while the Oxygen will be exhausted by the combustion, the light will be extinguished, and the water will rise into the jar,* showing that at least so much of the air is no longer present in its previous form.

---

* In order that nothing but Nitrogen may be left to exclude the water, phosphorus is the substance that must be burned. If an ordinary light be employed, the carbonic acid gas produced by its combustion retains much of the room of the oxygen. A few drops of Alcohol, lighted in a small floating cup, may be conveniently employed in this experiment.

C

i. Water consists of Oxygen and of another gas which is called Hydrogen, i. e., 'the producer of water'. Hydrogen, like Oxygen and Nitrogen, has neither taste nor smell; and a glass jar filled with it appears colourless.

j. In respect of combustion, Hydrogen is the opposite of Oxygen. A lighted taper plunged into Hydrogen is extinguished while the Hydrogen itself takes fire.

k. Hydrogen is the lightest of all known bodies. A balloon filled with it rises higher into the air than any bird can fly; and if the balloon is a large one it is capable of raising to that height a car containing several persons.

l. Chlorine is a gas which differs from the other simple gases in several respects. Its colour is greenish yellow. Its odour is extremely pungent and irritating, and its taste is analogous.

m. Several substances inflame in it spontaneously. Such are copper, in fine leaves; phosphorus; mercury, &c.

n. It destroys all animal and vegetable colours, and hence it is very useful in bleaching. It also destroys the volatile products of decaying animal or vegetable substances, thus removing the offensive odour and other bad effects which are apt to arise from such a source in hospitals, prisons, &c.

Aphorism XIII.

The Simple Non-metallic bodies which are not gaseous, are nine in number.

a. These are named Carbon, Boron, Silicon, Sulphur, Selenium, Phosphorus, Bromine, Iodine, and Fluorine.

b. Carbon (or charcoal) is the black substance which is left when wood is heated and the air excluded, or when the embers are extinguished by throwing water upon them.

c. Boron is a brown powder, which is a constituent of the substance called borax.

d. Silicon is a brown powder, which is a constituent of that hard substance, which strikes fire with steel, called flint (—or, iu Latin, silex).

e. Sulphur is a well known inflammable yellow substance.

f. Selenium is a rare substance which in many respects resembles sulphur.

g. Phosphorus is a substance which in appearance resembles wax. It shines in the dark; and a degree of heat less than that of boiling water is sufficient to inflame it. As it takes fire on being rubbed, it is employed in making the matches called lucifer matches.

h. Bromine is a dark red fluid with a very disagreeable smell (—as its name implies).

i. Iodine is a substance of a bluish black colour with a metallic lustre. Its smell resembles that of chlorine. When heated, it assumes the form of a vapour, which (as the name implies) is of a fine violet colour.

j. The name of Fluorine is given to a supposed element in a mineral called Fluor spar.

Aphorism XIV.

The Metals are forty-six in number.

a. Some of these, as Gold, Silver, Copper, and Lead, are familiarly known. Of the remainder, some are very rarely met with, and need not be here described. We shall describe at present only those of the metals which it is desirable that the reader should become acquainted with.

b. Platinum is the heaviest of the metals, being heavier even than Gold. It is white like Silver.

c. Mercury differs from the other metals in being fluid at ordinary temperatures. In the cold air near the north pole it becomes solid, and it may then be hammered into leaves like iron or copper. When strongly heated, it rises in vapour.

d. Tin is a metal which resembles silver in colour and lead in softness.

e. Zinc is a metal with a leaden colour. It takes fire in the air when heated red-hot, and floats in the air like flocks of wool.

f. Potassium is a metal which is so light that it swims upon water, where, at the same time, it inflames spontaneously.

g. Sodium also floats when thrown upon water, but it does not, like Potassium, inflame.

h. Calcium is a metal which is the basis of lime.

i. Of the simple substances which have been now partly described, all the bodies with which we are acquainted on the earth's surface are made up.

j. If any substance, such as salt or chalk, is made up of several simple substances, it is desirable that it should have a name suggestive of its component elements.

Aphorism XV.

In chemical enquiries let compound substances receive names suggestive of their component elements.

a. For example,—common salt is a compound of Chlorine and Sodium, and its chemical name is Chloride of Sodium.

b. Beginning with Oxygen, we shall now notice how the several Simple Bodies enter into compounds.

c. We have already stated that the atmosphere consists mainly of one part of Oxygen mingled with four parts of Nitrogen. The product of this mixture is not reckoned a chemical compound, because, as has been already explained, chemical composition is considered to have taken place, only when the qualities of the product are not what could have been predicted by taking together the properties of the separate ingredients.

d. Of the chemical compounds of Oxygen with Nitrogen there are five. The five compounds, in the order in which we shall take them, differ from one another in having each a larger proportion of Oxygen than the one before it in the list.

e. The first compound of Oxygen with Nitrogen is a gas which if a man breathes he laughs violently, and becomes for a time like one deranged.

f. The second compound of Oxygen with Nitrogen is a colourless gas, which, when allowed to mingle with the atmosphere, attracts an additional quantity of Oxygen and becomes red.

g. Charcoal and Phosphorus burn brilliantly in the second compound of Oxygen and Nitrogen; but this gas cannot be breathed, because if it were to mingle with the common air in the lungs, it would form the red gas already mentioned which is poisonous.

h. The third compound of Oxygen and Nitrogen, which is procured with difficulty, is an acid which does not here require special notice.

i. The fourth compound of Oxygen and Nitrogen is the red gas already mentioned. Water absorbs it readily.

j. The fifth compound of Oxygen and Nitrogen cannot be procured separately. In combination with water it is an acid which corrodes the flesh, and dissolves iron &c.

k. We have said [—see § 14. *j.*—] that it is desirable that substances should, in chemical enquiries, be denoted by names suggestive of their component elements ; but, from what has now been said of the compounds of Oxygen and Nitrogen, it must be obvious that the name ought to suggest not merely the elements but the relatively greater or lesser proportion of the elements in each compound. For example, the name of the second compound of Oxygen and Nitrogen ought to suggest its containing a greater proportion of Oxygen than the first compound. How this is effected, we proceed to explain,—premising some remarks on the various habits of various bodies in regard to their combining.

Aphorism XVI.

In the union of some bodies no limitation is observed with regard to the relative proportions in which they unite.

a. For example,—with a certain quantity of milk, a small quantity of water, or a large quantity of water, mingles equally well.

Aphorism XVII.

Some bodies combine in all proportions as far as a certain point, beyond which, combination no longer takes place.

a. Thus a certain quantity of water will dissolve common salt until a certain quantity has been dissolved, after which all the salt added remains undissolved. [This kind of limitation is called Saturation].

b. Since, in the two kinds of combination just described, the sensible qualities belonging to the combining substances in their separate state, either continue to be apparent in the compound which they form, or are not permanently altered, such combination is not strictly termed *chemical*: [—see § VIII. *ab.* and *ac.*]

Aphorism XVIII.

Bodies which unite chemically do so in certain fixed or definite proportions; and when bodies unite chemically in more than one proportion, the quantity of one element in each compound is found to be a simple of its quantity in the first.

a. Thus in the five compounds of Oxygen with Nitrogen above-mentioned, the proportions are those of 14 parts of Nitrogen to 8, 16, 24, 32, and 40 parts of Oxygen respectively.

Aphorism XIX.

The proportions in which two elements combine with *Hydrogen* indicate the proportions in which those two elements combine also with each other.

a. For example, 35 parts of Chlorine combine with one of Hydrogen [to form Hydrochloric acid], and 8 parts of Oxygen combine with one of Hydrogen [to form Water] ; and the compounds of Chlorine with Oxygen contain 35 parts of chlorine combined with 8, 32, 40, and 56 parts respectively of Oxygen. Now the numbers 8, 32, 40, and 56, are not successive simple multiples of 8, for these would be 8, 16, 24, 32, 40, 48, 56; therefore the law under consideration [—which is called the law of Definite Proportions] suggests the probability that there are at least three different compounds of Oxygen and Chlorine yet undiscovered.

b. Having ascertained that substances combine chemically in fixed proportions, chemists have inferred that the elementary bodies are not, like space, infinitely divisible, but that they consist of minute indivisible particles,—called Atoms.

c. It is supposed that the Atoms, though exceedingly small, are not devoid of magnitude [—as is supposed in the Nyáya;— see § XL. *b.* Book II].

d. It is also supposed that the atoms of any one element,— say of Gold—are alike in weight, but that they differ in weight from the atoms of other elements.

d. When bodies combine in only one proportion, it is supposed that the combination is of one atom of the one with one atom of the other; and when they combine in more proportions than one, that the combination is of one atom of the one with one, two, three, four, &c. of the other.

e. From this it follows that the numbers expressing the combining proportion express also the relative weights of the atoms of the combining elements.

f. This will appear clearly if we consider that if eight pounds of one element combine with one pound of another element so as to form nine pounds of a compound substance, in which each atom of the one element is joined to an atom of the other element, then an atom of the one element must be just eight times as heavy as an atom of the other.

g. The following is a table of the comparative atomic weights (or chemical equivalents) of the most important elements, with their symbols.

Oxygen 8.	O	Copper 32.	Cu
Nitrogen 14.	N	Iron 28.	Fe
Hydrogen 1.	H	Lead 104.	Pb
Chlorine 35.	Cl	Platinum 98.	Pt
Carbon 6.	C	Mercury 100.	Hg
Boron 11.	B	Zinc 32.	Zn
Sulphur 16.	S	Potassium 40.	K
Phosphorus 32.	Ph	Sodium 23.	Na
Gold 98.	Au*	Calcium 20.	Ca
Silver 110.	Ag		

h. When treating of the chemical composition of bodies it is found convenient to denote the elements by short symbols, as in Algebra:—[see § 3. *c* B. III]. Among the symbols in use are those given in the above table.

i. Each symbol is understood to stand for as many parts, by weight, of the element, as there are units in the number denoting its atomic weight or combining proportion. Thus O stands for 8 parts, by weight, of Oxygen; N for 14 parts of Nitrogen; H for 1 part of Hydrogen; and so on.

j. By combining the symbols of the elements we obtain the symbols of their compounds. Thus since the symbol Na stands for 23 parts of Sodium, and Cl for 35 parts of Chlorine, the symbol Na+Cl stands for 58 parts of Chloride of Sodium (or common salt—see § 15 *a*.) And this number 58 represents the atomic weight of Chloride of Sodium. The like remark applies to the atomic weights of all compounds.

---

* From the Latin *aurum*,—and so of the others.

D

k. Having determined thus much in regard to the principles of combination, let us revert to the five compounds of Oxygen with Nitrogen [§ 15. *d*]. From what is remarked under Aph. 18 and 19 *d.*, it will be understood how the symbolic formulæ and the atomic numerals of the five compounds of Oxygen and Nitrogen are as follows.

1st compound	N+O	....	22
2nd —	N+2O	....	30
3rd —	N+3O	....	38
4th —	N+4O	....	46
5th —	N+5O	....	54

l. An inspection of these symbols will show that, in these five compounds there is such a difference in the relative proportions of the elements as will require [see § 15. *k*] to be suggested by the chemical name of the compound. But there is another consideration which prevents us from giving to the five a set of names suggestive of nothing more than this distinction. For example, some of the compounds of Oxygen with Nitrogen are *acid*, and others are not. This difference being, for reasons which will appear in the sequel, a very important one, it requires to be recorded in the name of the compound ; and, with reference to this, we propound the following aphorism.

Aphorism XX.

Compounds of the elements ought to have names suggestive of the fact that they are *acid* or otherwise.

a. Of the five compounds of Oxygen and Nitrogen three are *not* acid, and these, accordingly, are named, simply with reference to their having respectively the first, the second, or the third and greatest proportion of Oxygen, as follows.

The Protoxide of Nitrogen.
The Deutoxide of Nitrogen.
The Peroxide of Nitrogen.

b. Of the compounds of Oxygen and Nitrogen the remaining two are acids. Let us consider how these are to be distinguished.

c. When any substance, by combining with two different proportions of Oxygen, is capable of forming two distinct acids, that acid which contains the larger proportion of Oxygen is distinguished by the syllable *ic* at the termination of its name, and the other, containing the lesser proportion of Oxygen, is distinguished by the termination *ous.* Accordingly the third and fifth compounds of Nitrogen with Oxygen, which are acids, are respectively named Nitrous acid and Nitric acid. The names of the five compounds of Oxygen with Nitrogen are therefore as follows.

1 Protoxide of Nitrogen.
2 Deutoxide of Nitrogen.
3 Nitrous acid.
4 Peroxide of Nitrogen.
5 Nitric acid.

d. Having considered the compounds of the two first elements with one another, we now turn to the third element —viz.—Hydrogen—and enquire what compounds it forms with the two preceding elements.

e. Hydrogen and Oxygen, atom to atom, produce water. The proofs of this fact will be given hereafter. The symbol of water accordingly is H+O, and its atomic numeral 9.

f. Hydrogen combining with Nitrogen produces Ammonia, a gas with a very pungent odour. This gas eagerly combines with

water.   In Ammonia there are three atoms of Hydrogen to one
of Nitrogen, so that its symbol is 3 H+N, and its atomic numeral 17.

g. Chlorine, the fourth in our list, combines with Oxygen,
with Nitrogen, and with Hydrogen, forming a variety of compounds, among which we shall notice at present only that with
Hydrogen. This Compound is a pungent acid gas, termed Hydrochloric Acid, which unites eagerly with water.   Its symbol
is Cl+H, and its atomic weight 36.

h. We have next to consider the combination of the second
class of elements with the first.

i. Carbon, in combination with Oxygen, produces an acid gas,
Carbonic Acid. This gas is fatal to animal life.   It is so heavy
that it may be poured from one vessel into another like water.
Hence, in some wells and caverns where it is formed naturally,
it occupies the lower parts, whilst the upper parts are free from
it ; and in the same place, therefore, where a dog is killed by it, a
man, from his superior height, is unhurt. The symbol of Carbonic
Acid is C+2 O, there being two atoms of Oxygen to one of Carbon.   Its atomic weight is 22.

j. When a diamond is burned in Oxygen gas, the result is
Carbonic Acid. This shows that the diamond consists of carbon.

k. Carbon with Nitrogen produces an inflammable gas with a
penetrating and peculiar smell,—the Carburet of Nitrogen.

l. With Hydrogen Carbon forms an inflammable gas, carbu-
retted Hydrogen—which may frequently be obtained from stagnant pools by stirring the mud. This gas in some places, as at
Jwálámukhí in the Panjáb, issues from the ground in such quan-

tities that it can supply a continual flame. A gas of this kind may be extracted from coal. In Europe it is employed to light up the streets.

m. Sulphur, combined with three atoms of Oxygen, produces Sulphuric Acid, a gas which readily unites with water. Its symbol is $S+3$ O, and its atomic weight 40.

n. Phosphorus, combined with five atoms of Oxygen, forms Phosphoric Acid. Its symbol is $Ph+5$ O, and its atomic weight 72.

o. Phosphorus combined with Hydrogen produces a gas which inflames spontaneously in the air. It is named Phosphuretted Hydrogen.

p. We have now to consider the combination of the Metals, the elements of the third class, with those of the other classes and with one another.

q. The combinations of most metals with Oxygen are commonly termed *rusts*. Gold and Silver have very little disposition to combine with Oxygen. They do not *rust*; and this is one reason why they are well adapted for making coins &c.

r. The Common red Oxide, or rust, of Iron consists of two atoms of Iron to three of Oxygen. This ratio being that of 1 to $1\frac{1}{2}$, the red rust of iron is called the Sesqui-Oxide* of Iron.—Its symbol is $2\ Fe+3$ O, and its atomic weight 40. There is a Protoxide of Iron, the symbol of which is $Fe+O$, and the atomic weight 36. The sparks that fly off from a mass of hot iron, when hammered by the blacksmith, consist of this Oxide.

---

* From the Latin prefix *Sesqui*, signifying half as much more.

s. Of Lead the Protoxide (litharge) is yellow, and the Deutoxide (minium) is red.

t. The Peroxide of Mercury is of a red colour.

u. With Chlorine Mercury forms two compounds. The Protochloride of Mercury (calomel), a tasteless white powder, is a valuable medicine. The Perchloride of Mercury (corrosive sublimate) is highly poisonous.

v. With Sulphur, as already mentioned, [see § 8. *ab*], Mercury forms a brilliant red powder—the Bisulphuret of Mercury (vermilion).

w. Mercury combines readily with several of the metals. Combined with gold it is employed in gilding vessels of silver.

x. The Oxide of Potassium, called Potassa, is a white substance with an intensely acrid (alkaline) taste. It corrodes the flesh. Its symbol is K+O, and its atomic weight 48.

y. The Oxide of Sodium, called Soda, very much resembles that of Potassium. Its symbol is Na+O, and its atomic weight 31.

z. With Chlorine, as already mentioned [see § 15. *a.*] Sodium forms common salt. The symbol of Chloride of Sodium [common salt] is Na+Cl, and the atomic weight 58.

aa. The Oxide of Calcium is commonly called lime. Its symbol is Ca+O, and its atomic weight 28.

ab. Having now considered the *binary* compounds,—those into which only two elements enter,—we have next to consider those compounds into which a greater number of elements enter. In *Aph.* 20 the distinction between Acids and other compounds has been adverted to. One distinction which belongs to most acids, as is implied in the name, is a sour taste. Another characteristic, or *test*, is this, that they change a vegetable blue infusion, such as that of the Hibiscus flower, to red. Among the binary compounds there are others which, like Potassa, have an acrid taste [and which are termed *alkalis*], which restore the blue colour to the infusion, or which go further and change it to green With these alkalis, as well as with other oxides, the acids are disposed to form further compounds, in regard to which we remark as follows.

Aphorism XXI.

Acids combine with Oxides to form Salts.

a. It has already been mentioned that common salt is a compound of Chlorine and Sodium ; so that it is not now asserted that *all* salts are compounds of acids and oxides, but only that very many are so ; and of these we now proceed to give some account.

b. Acids as already mentioned [—see § 20. *c.*] have names ending in *ic* or *ous.* In the name of the salt the *ic* is changed to to *ate,* and the *ous* to *ite.* Thus a salt into which Nitric Acid enters is called a Nitrate, and one into which Nitrous Acid enters, a Nitrite.

c. As it is always with the *Oxides* of the metals that the Acids enter into combination, it is not necessary to notice this in naming the salt, unless when more than one oxide of the same metal combines with acids. Thus, instead of Nitrate of the Oxide of Silver, it is sufficient to say Nitrate of Silver, but we must distinguish between the Sulphate of the Protoxide of Iron, and the Sulphate of the Sesquioxide of Iron.

d. In the following list, the Acid and Oxide in each case is mentioned in the left hand column, with the respective symbols and atomic weights, and the resulting symbol and atomic weight in the right-hand column.

$$\left. \begin{array}{l} \text{Nitric Acid} = N + 5O = 54. \\ \text{Potassa} = K + O = 48. \end{array} \right\} \begin{array}{l} \text{Nitrate of Potassa} \\ = KO + N\ 5O = 102. \end{array}$$

Nitrate of Potassa is the well-known salt, commonly called Nitre or Saltpetre, which is employed in the manufacture of gunpowder.

$$\left. \begin{array}{l} \text{Nitric Acid} = 54. \\ \text{Ammonia} = 3H + N = 17. \end{array} \right\} \begin{array}{l} \text{Nitrate of Ammo-} \\ \text{nia} = 71. \end{array}$$

Nitrate of Ammonia is employed in the preparation of the Protoxide of Nitrogen, the gas which intoxicates one who breathes it: [see § 15. e].

$$\left. \begin{array}{l} \text{Nitric Acid} = 54. \\ \text{Oxide of Silver} = Ag + O \\ = 118. \end{array} \right\} \begin{array}{l} \text{Nitrate of Silver} \\ = 172. \end{array}$$

This salt is employed by medical men as a caustic: (—lunar caustic).

$$\left. \begin{array}{l} \text{Sulphuric Acid} = S + 3O \\ = 40. \\ \text{Soda} = Na + O = 31. \end{array} \right\} \begin{array}{l} \text{Sulphate of Soda} \\ = NaO + S3O = 71. \end{array}$$

Sulphate of Soda is one of the salts employed in the galvanic battery.

$$\left. \begin{array}{l} \text{Sulphuric Acid} = 40. \\ \text{Lime} = Ca + O = 28. \end{array} \right\} \begin{array}{l} \text{Sulphate of Lime} \\ = 68. \end{array}$$

Sulphate of Lime (—commonly called Plaster of Paris—) is white like chalk. It is much employed in making statues by means of a mould.

$$\left.\begin{array}{l}\text{Sulphuric Acid}=40.\\ \text{Oxide of Copper}=Cu+O\\ =40.\end{array}\right\}\begin{array}{l}\text{Sulphate of Copper}\\ =80.\end{array}$$

Sulphate of Copper (—commonly called blue vitriol—) is a well-known salt which is employed in the galvanic battery.

$$\left.\begin{array}{l}\text{Sulphuric Acid}=40.\\ \text{Protoxide of Iron}=Fe+O\\ =36.\end{array}\right\}\begin{array}{l}\text{Sulphate of the Pro-}\\ \text{toxide of Iron}=76.\end{array}$$

This salt, being of a green colour, is commonly called green vitriol.

$$\left.\begin{array}{l}\text{Hydrochloric Acid}=36.\\ \text{Ammonia}=17.\end{array}\right\}\begin{array}{l}\text{Hydrochlorate of}\\ \text{Ammonia}=53.\end{array}$$

This salt, commonly known as Sal Ammoniac, is employed in cooling water, and for many other purposes.

$$\left.\begin{array}{l}\text{Carbonic Acid}=C+2\ O\\ =22.\\ \text{Lime}=28.\end{array}\right\}\begin{array}{l}\text{Carbonate of Lime}\\ =50.\end{array}$$

The substance commonly known as Chalk, is a Carbonate of Lime.

$$\left.\begin{array}{l}\text{Boracic Acid}=B+6\ O\\ =68.\\ \text{Soda}=31.\end{array}\right\}\begin{array}{l}\text{Borate of Soda}\\ =99.\end{array}$$

This salt is the substance well known as Borax.

$$\left.\begin{array}{l}\text{Phosphoric Acid}=Ph.\\ +5\ O=72.\\ \text{Lime}=28.\end{array}\right\}\begin{array}{l}\text{Phosphate of Lime}\\ =100.\end{array}$$

This salt is the principal ingredient in the bones of men and other animals.

E

e. Thus far we have been engaged in tracing the Elements through their successive combinations until we arrive at some of the well known compound substances which men daily make use of. The production of a substance by the combination of the proper ingredients is called *Synthesis.** The converse process, where we decompose a compound in order to determine what are its constituent elements, is called *Analysis.†*

Aphorism XXII.

The analysis of Chemical compounds is effected by Heat, by Electricity, or by the introduction of some substance having a stronger attraction for one of the substances united than they have for each other.

a. For example, the red Oxide of Mercury [§ 20. *t.*] when heated, is resolved into metallic Mercury and Oxygen gas. This furnishes one means of procuring Oxygen in a separate state. The method of conducting the operation will be explained in a treatise subsidiary to this Synopsis.

b. Water is decomposed by Electricity into Hydrogen and Oxygen.

c. Sulphate of Copper (the *tútiyá* or blue vitriol of the bázárs) may be decomposed by dipping a polished piece of iron into a solution of the salt. The acid, having a stronger attraction for the iron than for the copper, unites with the iron, and deposits the copper on its surface. When the whole of the copper

---

* From a Greek word signifying "to put together."
† From a Greek word signifying "to unloose."

has been deposited, the solution will be found to consist of Sulphate of Iron (the *kasís* or green vitriol of the bázárs).

d. The comparative attractions of various substances having been first ascertained by experiment, have been registered by chemical enquirers in tables, by referring to which we learn what substances to employ in decomposing any compound body.

e. The Atomic number of any substance is also called its *equivalent* number, because it indicates the proportional number of pounds or ounces of the substance which are exactly sufficient to supply the place of another substance in a compound. Thus, in the experiment cited above, if there were 31 ounces of copper in the solution of blue vitriol, it would be found that 28 ounces of iron were sufficient to supply the place of the copper ; or, in other words, that the acid which had dissolved 31 ounces of copper, could dissolve no more than 28 ounces of iron. The powers of saturation of iron and copper have thus been experimentally ascertained to be in the ratio of 28 to 31, as is recorded in the table [at § 19. *g.*] A knowledge of the equivalent numbers saves the chemist from all doubt as to the exact quantity of any substance which is necessary to effect any proposed composition or decomposition.

f. Substituting the symbols for the names of the Sulphate of Copper and of the Iron, the result of the decomposition may be concisely expressed as follows.

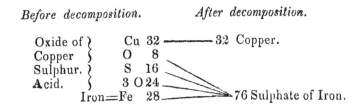

Before decomposition.   *After decomposition.*

Oxide of ⎱ Cu 32 ———— 32 Copper.
Copper ⎰   O   8
Sulphur. ⎱ S 16
Acid.   ⎰ 3 O 24
     Iron=Fe 28 ————⟶ 76 Sulphate of Iron.

g. It will be observed that, in accordance with the principle of Chemical Equivalents, the quantity of Oxygen that was required to make the 32 parts of Copper in the salt into an Oxide [see § XXI. *c.*], is just the quantity that is required to make the 28 parts of Iron into an Oxide. If the Iron, instead of being dipped into a solution of Sulphate of Copper, is immersed in a watery solution of Sulphuric Acid, then, in order to combine with the acid, it takes the necessary supply of Oxygen from the *water.* The consequence is that the Hydrogen of the water is set free, and comes away in the shape of bubbles. This decomposition may be illustrated as follows.

This furnishes one means of procuring Hydrogen in its separate state. How to collect it as it comes away will be explained in a subsidiary treatise.

h. In the foregoing examples one out of three substances, by taking possession of the second, turns the third adrift. In certain cases of decomposition there are four substances forming two compounds, which two, on meeting, make a mutual exchange of an ingredient. For example, let Hydrochlorate of Lime dissolved in water,* be mixed with a solution of Sulphate of Ammonia :—the Sulphuric Acid will combine with the Lime, and the

---

* The intelligent chemist will understand that the change of the chloride into a hydrochlorate, on solution in water, does not affect the principle to be exemplified, and is not therefore here adverted to.

Hydrochloric Acid, instead of flying off (as it would have done if Sulphuric Acid alone had been added) combines with the Ammonia which the Sulphuric Acid had deserted. This may be represented as follows,—the continuous lines showing the direction of the one pair of ingredients from left to right, and the dotted lines the direction of the other pair.

Before decomposition.   *After decomposition.*

Sulphate of ⎰ 3 H+N............................Hydrochlorate
Ammonia.   ⎱ S+3 O   of Ammonia.

Hydrochlor- ⎰ Ca+O   Sulphate
ate of Lime. ⎱ Cl+H   of Lime.

i. We have hitherto considered only such products of the elements as can be reproduced, by merely human means, after their elements have been separated. The red Oxide of Mercury, for example, may be separated into Mercury and Oxygen [§ 22. *a.*]; but we can very easily cause the Mercury to recombine with Oxygen and to reappear in the form of the red Oxide. It is otherwise, however, with certain other substances. We can decompose a piece of wood, and ascertain that it consists of Oxygen, Hydrogen, and Carbon; and we can decompose a piece of flesh, and ascertain Nitrogen, and Carbon; that it consists of Oxygen, Hydrogen, but we cannot, out of these elements, reproduce, in our laboratories, either wood or flesh. To the reproduction of these another principle is indispensable, viz. *life.*

Aphorism XXIII.

The principle of Life modifies chemical action.

a. A seed of a plant, placed in the earth, produces changes in the matter taken up into the plant which cannot be imitated by art. For example, the seed of the Madder developes a well known colour which can be derived from no other source; and the seed of the Mango produces a fruit the flavour of which is equally inimitable. Let us, then, now turn to the consideration of plants, the peculiar nature of which furnishes the province of Botanical science.

SECTION VIII.

BOTANY.

b. Plants are regarded first under the following division.

Aphorism XXIV.

Plants either have visible flowers, or they have not.

a. Among the plants which have no visible flowers are the mushrooms. Plants with visible flowers are such as the Mango-tree, &c.

b. The *Ficus religiosa* and other fig-trees are not to be reckoned among plants with invisible flowers, because their flowers, inside of the fruit, are visible by the aid of a microscope.

c. Of the thread-like fibres seen in flowers, some called the *stamens*, are male; the others called the *pistils*, are female. Those on the top of which there is a kind of powder or dust, are the stamens; the others are the pistils.

d. Flowering plants are first classified according to the number of their stamens. These classes are again sub-divided according to the number of the pistils; and the species of the plant is determined by the form of the leaves and other characteristics

e. When the dust from the stamens falls upon the pistil, then the plant bears seed, but not otherwise. Hence, if the stamens of the blossom be removed, no fruit will be produced. But if, after the removal of the stamens from a flower, the dust from the stamens of a plant of a kindred species be sprinkled over the pistil, then the seeds produced will partake of the qualities of both. By this means new varieties of flowers have been produced.

f. In some kinds of trees the flowers on one tree contain stamens only, and on another pistils only. The former may be called male trees, the latter female. The male trees bear no fruit, and are commonly called barren. The reason of their barrenness is obvious from what has been said. The female trees bear fruit when they are planted near the males, as in the case of the date tree and the like.

g. The species of known trees are reckoned at about 50,000; so that our limits preclude any particular description of them. Of the structure of plants we shall treat further under another head—that of Organography.

h. Plants, though they have life, have not the power of locomotion. This distinguishes then from animals. The science that treats of animals is termed Zoology.

SECTION IX.

ZOOLOGY.

i. Animals are regarded first under the following division.

Aphorism XXV.

The animal kingdom is divided into four provinces—viz. the *Vertebrata,* the *Mollusca,* the *Articulata,* and the *Radiata.*

a. The first division, or sub-kingdom, that of the *Vertebrata,* consists of the animals possessing a spine, as man. The second division, that of the *Mollusca,* consists of animals with very soft bodies, many of which have shells, as the snails. The third division, that of the *Articulata,* consists of animals with jointed bodies, like the centipede. The fourth division, that of the *Radiata,* consists of animals whose members are symmetrically disposed around a centre, as in the starfish, the sponge, and the *animalcula,* discernible only by the aid of the microscope, which swarm in stagnant water.

b. Of the first division, the *Vertebrata,* there are four subdivisions or classes. The first is that of the *Mammalia,* or animals which give suck; the second, of the *Aves,* or Birds; the third, of the *Reptilia,* or Reptiles; and the fourth, of the *Pisces,* or *Fishes.*

c. Of the *Mammalia,* again, there are twelve sub-divisions or orders. The order *Bimana,* or two-handed, includes man alone; the *Quadrumana,* or four-handed, are the monkeys; the *Cheiroptera,* or hand-winged, are the bats. The *Insectivora,* or eaters of insects, include the musk-rat; the *Carnivora,* or eaters of flesh, the tiger; the *Cetacea,* or animals like the whale, the porpoise of the Ganges; the *Pachydermata,* or animals with thick skins, the elephant; the *Ruminantia,* or those that chew the cud, the ox; the *Edentata,* or animals with few or no teeth, the pangolin; the *Rodentia,* or gnawing animals, the mouse; the *Marsupiata,* or animals with a pouch, the Kangaroo; and the *Monotremata,* the Ornithorhyncus, or duck-billed quadruped of Australia.

d. Like the *Vertebrata,* the three other great divisions of the animal kingdom contain many sub-divisions which we do not propose here to detail.

e. Let us now advert to the structure of plants and animals ;— and first of plants.

SECTION X.

ORGANOGRAPHY AND VEGETABLE PHYSIOLOGY.

f. We have spoken of plants and animals as differing from other masses in the circumstance of having *bodies.* A body is defined by Gautama [Book I. Aph. XI.] as the site of organic action, of organs, and of sentiments. It is in consideration of their organic action, and of their having organs appropriate thereto, that we speak of *plants* as having bodies [—B. I. Aph. XI. *d*]. We have no certain evidence that plants have sensations, or senti‐ ments. Let us now consider the order in which a plant is developed from a seed.

Aphorism XXVI.

A seed, being placed in the ground, in due time sends forth two shoots, one ascending and forming the stem, and the other descending and forming the root.

a. When the shoot appears above ground it shows a strong desire—if we may so term it—for light. If the light be excluded, the plant languishes, and the leaves do not acquire their green colour. If light be admitted by a small opening, the plant will incline towards the opening.

Aphorism XXVII.

The food of plants is derived from the earth by the root, and from the air by the leaves.

F

a. It has been already mentioned that the substance of plants consists chiefly of Oxygen, Hydrogen, and Carbon. Oxygen and Hydrogen may be supplied by water, and every one knows how necessary it is for plants to be furnished with water, which they take in by their roots.

b. From the air plants derive Carbon. Carbon exists in the air, in combination with Oxygen, in the shape of Carbonic Acid gas. This gas is produced in large quantities by the breathing of animals, as well as from other sources.

c. To show that Carbonic Acid exists in the air exhaled from the lungs, breathe, through a tube, into a transparent solution of Lime in water. The lime-water will become turbid, the lime being converted into Carbonate of Lime, or chalk, which is insoluble in water.

Aphorism XXVIII.

The leaves perform for the plant the functions of the stomach and lungs of animals.

a. The stomach of an animal separates the nutritious portions from the food, and rejects the remainder. The leaves of plants do the same. Thus, Carbon being required in a solid shape to form the stem and branches of the plant, it is separated from the Carbonic Acid by the leaves, and the Oxygen of the Carbonic Acid is rejected.

b. So we find that animals exhale a gas which, in a large quantity, is poisonous to animals, but which is necessary to plants; and plants in return inhale this gas, and exhale the Oxygen which is indispensable to animals.

c. We have seen the proof that Carbonic Acid gas is exhaled by animals. Let us now see the proof that Oxygen gas is exhaled by plants.

d. Place a handful of fresh leaves under a large glass vessel quite filled with water and inverted in another vessel of water. Expose the whole to the light of the sun, and the leaves will give out Oxygen, which may be recognised by its chemical properties already described.

e. Of the structure of the flowers of plants, and of the functions of the parts, some account has already been given when treating of that classification of plants (the Linnæan) which is founded on the diversities in the number and arrangement of the stamens and pistils.

f. Let us now consider the structure of the bodies of animals, —and first of the human body.

SECTION XI.

ANATOMY AND PHYSIOLOGY.

Aphorism XXIX.

The parts of the human body are the skeleton, the muscles, the brain, the stomach, the heart, the lungs, the bloodvessels, &c.

a. The skeleton consists of 254 bones. Of these an account will be given in a subsidiary treatise.

b. The muscles constitute the bulk of what is commonly called the flesh. At the impulse of the will these are shortened or lengthened, and thus they move the bones of the skeleton to which they are attached.

c. The mass of the brain is located in the scull. Its substance, extending downwards through the spinal column, is distributed in small branches, called nerves, to most parts of the body. Where these do not extend, as into the hair or the nails, there is neither sensation nor voluntary motion.

d. The food eaten is divided by the stomach into what is useful and what is not.

e. The digested food [in the shape in which it is called chyle] is carried to the heart by a channel [called the left subclavian vein]. Here it mixes with, and reinforces, the blood, which is the medium by which all the parts of the body, solid or fluid, are supplied with nourishment.

f. From the heart, by its constant action, the blood is forced, through the arteries, into the lungs, and afterwards throughout the body,—returning to the heart through the veins.

g. In the lungs the blood absorbs Oxygen, and parts with Carbon, as has been already noticed when treating of plants,—which, on the contrary, absorb Carbon and part with Oxygen.

h. Having in some manner described the separate portions which constitute the globe, so far as we are acquainted with it, let us advert to the distribution of these constituent portions.

SECTION XII.

GEOGNOSY AND MINERALOGY.

i. Under the head of Geography some account has been given of the arrangement of the earth's surface,—its mountains its rivers, its seas, &c. Now all mountains are not formed of the same materials. Some, as the Himálaya, are formed of very hard rock; and others, as the sandstone range parallel to the Himálaya [—see B. II. Aph. 32.] of softer materials. In some kinds of rocks are found diamonds, &c.; and such useful substances as coal are never found except in company with certain other kinds of rocks. It is obviously desirable both to be able to recognise useful minerals when we see them (—which it is the business of Mineralogy to teach us)—and to know the general arrangement of the mineral masses of the earth, so that we may know where to seek for such minerals as we require, (—and this knowledge constitutes Geognosy).

Aphorism XXX.

Minerals may be divided into two classes, the crystallized and the uncrystallized.

a. The diamond, the topaz, and the emerald, are examples of crystallized minerals. *Kankar,* or nodular limestone, and clay, are examples of uncrystallized minerals.

b. Minerals, when crystallized, can frequently be recognized by mineralogists merely from the form of the crystals.

c. In the case of uncrystallized minerals, we must judge by means of colour, weight, and other characteristics; but especially by chemical analysis.

Aphorism XXXI.

As an onion is enveloped in many successive coats, so the crust of the earth consists of several successive layers of minerals; but in most places these layers, or *strata*, are rent and disarranged.

a. For example,—in some places the country consists of layers of chalk, &c., level like the mud deposited at the bottom of a pond. In other places these layers appear disarranged and rent by hard rocks, which rise up in the form of mountains. How this has come to pass we shall enquire in the next section [—on Geology].

b. The practical utility of Geognosy is very great. For example, we remarked that the arrangement of the strata of the earth bore a general resemblance to that of the coats of an onion. Now enquirers, by examining the structure of the rocks composing these successive layers, and the fossil remains which belong exclusively to each, have been able to determine the order of superposition which prevails over the whole world, so far as the matter has been yet investigated.

c. If therefore one versed in such enquiries finds, at the surface of the earth in any country, a particular layer, which, to his knowledge, always lies beneath the coal-formation, he can warn the miner not to expend his labour fruitlessly in seeking for coal there.

d. Again, if he find those kinds of rock underneath which coal is generally found, he may encourage the miner to persevere, when he sees him inclined to give up the search in despair. And so with the other useful substances which the miner obtains from the earth.

e. The proofs of all these assertions cannot be given in so slight a sketch as this; but their verification or refutation would seem an object not unworthy of the labour of the intelligent.

f. It would appear that on various parts of the earth where there is now dry land, the sea formerly flowed; and probably many places are submerged, where there was formerly dry land. No such animals and trees now exist on the face of the earth, as are known to have existed in primeval times.

g. Many of those ancient animals were of very strange kinds. Anatomists can sometimes recognise the species of an animal by means of a single bone; how much more so then, when entire skeletons have been discovered. We shall describe very briefly some of the animals which, from an examination of their skeletons embedded in rocks, have been ascertained to have existed in this globe at the period when those rocks were formed.

h. The Mammoth, the elephant of the ancient world, was about 9 feet high, and above 16 feet long from the point of the nose to the end of the tail. Its tusks were nine feet and a half in length, and were very much curved. Its body was covered with hair.

i. The Megalosaurus, or alligator of the ancient world, was between 40 and 50 feet in length. Its bones are found in England, a country which is now too cold for alligators.

j. The Iguanodon, or gigantic iguana, a kind of long-tailed lizard, grew to above 70 feet long.

k. The Ichthyosaurus, or fish-lizard, must have resembled an alligator with fins instead of feet. The largest species was about 20 feet long. Its eyes were larger than a man's head.

l. The Plesiosaurus differed from the Ichthyosaurus in having a very small head, and a long neck like the body of a serpent.

m. The Pterodactyle, or flying lizard, had a head like a crocodile, and wings like a bat.

n. Now, how did the arrangement of the materials of the earth, which we have been describing [—see Aph. 24 B. II], come to take place? This we proceed to consider—[under the head of Geology—i. e. the *rationale* of the Earth's being as it is].

SECTION XIII.

GEOLOGY.

Aphorism XXXII.

The interior of the earth is probably a mass of ignited matter.

a. If all the earth's heat were derived from the sun, it might be expected that the surface should be the warmest, and that as we dug into the earth we should come to colder portions. But the lower we descend in mines, the higher is the temperature.

b. In many places there are burning mountains, or volcanoes, which are produced as follows. From the interior of the earth a mass of melted mineral matter makes its way upwards through the upper strata, which it ruptures and forces out of their previously horizontal position. Issuing from the opening thus made, it forms a mountain.

c. In the course of time, clouds assemble on the top of the mountain, and thus streams are produced.

d. By the action of those streams the substance of the mountain is abraded, and quantities of it in powder suspended in the water are carried towards the ocean. Where the current, on meeting the ocean, is stayed, there the matter which the waters in motion had kept suspended, is deposited; and an island is gradually formed, which divides the mouth of the river into two.

e. In each of these the same process is repeated. An island formed in each mouth again doubles the number of mouths, and thus the mouths of the river sometimes become very numerous.

f. It was in this way that the hundred mouths of the Ganges were forme d, and the adjoining region called the *Sundarbans.*

g. In each of the islands thus produced, many trees spring up, and the bones of animals accumulate there, whilst numerous shells may be found along their shores.

h. If a volcano should now break forth in such a place, the region would be elevated, and shells and other marine productions would be found on the high grounds, as they are found on the tops of many mountains at this day.

i. It is in no short space of time that all this takes place; therefore man, whose days are few, cannot observe the process from first to last; but as, after viewing trees of all sizes in a forest, we feel sure that the large trees once resembled the little ones, and that the little ones are imperceptibly growing up to be like the large ones, so, by observing such processes on a small scale, as at the mouths of small streams, where small islands, containing various things floated down are formed and ever altering, we come to the conclusion that similar geological phenomena on a far larger scale must be due to similar causes.

j As the strata of the Sundarbans, if elevated by subterraneous fire, would be found to contain many bones and complete skeletons of tigers, deer, crocodiles, and other such animals as had been accustomed to frequent that region, so, in the strata which have been formed in other places in very ancient times, we might expect to find the remains of the animals that then lived. And this is the case; as was shown in the preceding section.

k. As the matters that we have been considering are not all ascertainable by the evidence of Sense alone, it is time that we turn to the consideration of the second species of evidence,— viz. that of Inference.

॥ श्री परमेश्वरो जयति ॥

। अथ द्वितीयाध्यायः ।

—oo—

सङ्ख्यापरिमाणयोस्तदाश्रयद्रव्यविचारनैरपेक्ष्येणेह विचारः करिष्यते । अत्र हेतुः पूर्वप्रकरणान्ते प्रदर्शितः ॥

अत्र निर्दिष्टसंख्याविचारविषयं शास्त्रं व्यक्तगणितमित्युच्यते ॥

दृश्यमान्तसंख्याव्यवहारबीजमाह ॥

संख्या दशावधि वर्धन्ते मनु-
ष्याणां दशाङ्गुलित्वात् ॥ १ ॥

क

यदा सजातीयानां समूहो विलोक्यते यथा मनुष्या-
णां तत्संख्याया: स्मरणं चिकीर्ष्यते चेत्तर्हि तत्र सङ्के-
तार्थमङ्गुली: प्रयोजयेत् ।   एकैकमनुष्यायैकैकामङ्गु-
लीमुत्थाप्य तस्या: संज्ञा क्रमेण एकद्ध्यादिकां कुर्यात् ।
अन्तिमाङ्गुल्युत्थापनानन्तरं पुन: प्रथमाङ्गुल्युत्थापनप्र-
सङ्ग: स्यात् सच यावद्द्वारं भवेत् तज्ज्ञापनायोपाय: प्र-
कल्प्य: ।   रूढस्य संख्याकरणप्रकारस्योत्पत्ति: कदा-
चिदेवमेव स्यात् यतो ऽत्राङ्गुष्टद्विर्दशावधिर्भवति ।   य-
दि मनुष्याङ्गुल्यो द्वादश भवेयुस्तर्ह्यङ्गुष्टद्विर्द्वादशा-
वधिर्भवेत् ॥

गणितसम्बन्धिनो विधयो बापूदेवकृतगणितप्रकार-
णादवगन्तव्या: ॥

यदातिमहत्यो: संख्ययोर्गुणनं भजनं वा विधेयं
तदा सामान्यरीत्या तत्सम्पादने महागौरवं क्लेशश्च भ-
वति अतस्तयो: संख्ययो: तथा अन्ये संख्ये कल्प्येते याभ्यां
फलं अल्पक्लेशेन लघुप्रकारेण ज्ञायते येच घात-

प्रमापकसंज्ञे स्याताम् । अथ घातप्रमापकाख्यसंख्या-
विशेषाः ॥

ययोः संख्ययोर्गुणनफलं भजनफलं वा ज्ञा-
तव्यं तयोर्घातप्रमापकयोर्योगादन्तराद्वा झ-
टिति सारण्यां गुणनफलं भजनफलं वा ल-
भ्यते एवमिष्टघाततन्मूले अपि ॥ २ ॥

यत्रोत्तरोत्तरमेकयैव संख्ययाधिकाः संख्या वर्त्तन्ते
यच्चाद्यसंख्या चयतुल्या स्यात् ताट्टग्येका संख्यानां श्रे-
णी कल्प्या । यथा । १ । २ । ३ । ४ । ५ । इ-
त्यादि । अत्र चयः । १ । ततो द्वितीयैका संख्यानां
श्रेणी तथा कल्प्या यथा तत्र संख्या उत्तरोत्तरं द्वि-
च्यादिगुणाः स्युर्यत्र चाद्यसंख्या गुणतुल्या स्यात् । यथा
। १० । १०० । १००० । १०००० । इत्यादि । अथेयं
श्रेणी प्रथमश्रेण्या अधस्तात् क्रमेण लिख्यते ॥

	१		२		३		४	
	१०		१००		१०००		१००००	
	५		६		७	
	१०००००		१००००००		१०००००००	

अचोर्ध्वपंक्तिस्थाः संख्याः क्रमेणाधस्थितसंख्यानां घातप्रमापकसंज्ञाः स्युः । तचेदमवगम्यते । तथाहि ॥

ऊर्ध्वपंक्त्यां द्वयोः संख्ययोर्योगो यत्र स्यात्तदधस्ता-द्द्वितीयपंक्त्यां स्थिता संख्या द्वितीयपंक्तिस्थयोः तद्योज्य-योजकाधरसंख्ययोर्वधेन समो भवतीति । यथा । ऊर्ध्वपंक्त्यां । २ । ४ । अनयोर्योगः । ६ । एतदधो द्वितीयपंक्त्यां वर्त्तमाना । १०००००० । इयं संख्या । २ । ४ । एतदधरयोर्द्वितीयपंक्तिस्थयोः । १०० । । १०००० । अनयोः संख्ययोर्वधेन समाना ॥

एवमेव उपरितनपंक्त्यां संख्ययोरन्तरं यत्र स्यात् त-

दधस्ताद्द्वितीयपंक्त्यां स्थिता संख्या द्वितीयपंक्तिस्थयाः तद्वियोज्यवियोजकाधरसंख्यया: लब्धा समा भवति य-था । ६ । ४ । अनयोरन्तरम् । २ । एतदधोधर-पंक्त्यां स्थिता । १०० । इयं संख्या । ६ । ४ । एतद्-धरयोः द्वितीयपंक्तिस्थयोः । १००००० । १०००० । अनयो: संख्ययो: लब्धा समा भवतीति ॥

एवमेवोपरितनी संख्या द्विचादैर्गुणिता भक्ता वा ऊर्ध्वपंक्त्यां यच्च भवेत् तदधःस्था संख्या तदुपरितनसं-ख्याधःस्थसंख्याया द्विचादिसंख्यापूर्वको घातो द्विचादि-संख्यापूर्वकघातमूलं वा स्यात् । यथा ऊर्ध्वपंक्त्यां । २ । इयं चिभिर्गुणिता । ६ । एतदधःस्था । १००००० । इयं संख्या । २ । एतदधःस्थाया । १०० । अस्या घनो भवतीति । अथैकाद्ययुतान्तसंख्यानां घातप्रमापका: सारण्यां लिख्यन्ते तेभ्यो यस्याः कस्याश्चिदपि संख्याया घातप्रमापको ल्पायासेन लभते ॥

लाघवार्थं चिह्नव्यवहार: ॥ ३ ॥

चिह्नव्यवहार इति । गणिते लाघवार्थं अनेकानि चिह्नानि व्यवहृतानि भवन्ति । अङ्कयोर्मध्ये लिखितं (+) इदं चिह्नं तद्योगज्ञापकं कल्पितम् । यथा २+४ इत्येतेन द्विचतुर्योगः क्रियतामिति ज्ञाप्यते ॥

तिर्यक्सरलरेखाद्वयं समत्वबोधकं कल्पितं । यथा २+४=६ इत्यच द्विचतुर्योगफलं षड्भिः सममिति बोध्यते ॥

संख्ययोर्मध्ये (−) एवं तिर्यग्रेखा आद्याया द्वितीया शोध्येति द्योतयति । यथा ६−२=४ इत्यनेन षड्भ्यो द्वयोः शोधितयोः शेषं चतुर्भिः सममिति द्योत्यते । अपिच षड्भ्यो द्वयस्य वर्जनेनैव चत्वारः शिष्यन्ते नान्य-संख्यावर्जनेनेत्यतोऽयं धर्मः सर्वसंख्याविषयको नास्ति । परन्तु कतिचिद्धर्माः सर्वसंख्याविषयका भवन्ति । य-था दौ पञ्चचानयोर्योगः सप्त । ७ । अन्तरंच चयम् । ३ । तद्योगान्तरयोर्योजनेन दशोत्पद्यन्ते एते नि-र्दिष्टसंख्ययोर्महत्या द्विगुणा भवन्ति । पुनस्तद्यो-

गादन्तरे शोधिते चतुष्कमवशिष्यते वृदमल्पसंख्याया
द्विगुणं भवति । वृद्धं सर्वंच विद्यात् ॥

तथाहि । द्वौ मुद्रासञ्चयौ वस्त्रबद्धौ कल्प्येतां यच
मुद्रासंख्या न ज्ञायते । तच वस्त्रबद्वसञ्चययोर्यो भेदः
तेन सहितः तयोर्योगो बहुमुद्रासञ्चयाद्विगुणो भवति
तेन च रहितस्तयोर्योगो ऽल्पमुद्रासञ्चयाद्विगुणो भव-
तीति निश्चीयते ॥

अथ बहुमुद्रासञ्चयश्चेत् (अ) कारेण द्योत्यते अ-
ल्पमुद्रासञ्चयश्च (क) कारेण तदेदं समीकरणरूपमु-
त्पद्यते ॥

$$(\text{अ}+\text{क})+(\text{अ}-\text{क})=२\,\text{अ}$$

$$(\text{अ}+\text{क})-(\text{अ}-\text{क})=२\,\text{क}$$

अच अ क अनयोरिष्टसंख्याभ्यामुत्थापितयोरपि
पच्चयोः साम्याविघात एव स्यात् ॥

एवमज्ञातसंख्यावगमकविधिकथनाय य ईप्सितो वर्ग-

व्यवहार: स बीजगणितसंज्ञको भवेत् ॥

बीजगणिते फलमव्यक्तम् ॥ ४ ॥

पूर्वोक्तचिह्नेभ्यो ऽन्यान्यपि क्रियाद्योतकचिह्नानि भ-
वन्ति । तथाहि संख्ययोर्मध्ये लिखितं (×) इदं चिह्नं
तयोर्वधं द्योतयति । यथा अ × ब इदं अकारद्योत्य-
संख्या बकारद्योत्यसंख्यया गुणनीयेति द्योतयति ।
अथाव्यवहितचिह्नौ सन्निहितवर्णाविपि खवधं द्योत-
यत: । यथा (अक) अच चेत् अ = २ क = ३ तदा
अक = ६ ॥

अथैकवर्णस्याधो रेखां कृत्वा तदधस्तादपरवर्णे लि-
खिते यत्सम्पद्यते तदुपरितनवर्णद्योत्यसंख्याया अधस्त-
नवर्णद्योत्यसंख्यया लब्धं द्योतयति । यथा यदि अ
= ६ क = २ तदा $\frac{अ}{क}$ = ३ ॥

संख्या खीयगुणा यथा अ × अ इयमेवं लिख्यते
$अ^२$ । इदं अकारस्य वर्ग उच्यते ॥

अथ यदा स्वगुणा संख्या पुनस्तयैव हन्यते यथा अ×
अ×अ तदिदं अ एवं लिख्यते । इदं अकारस्य
घन इत्युच्यते ॥

अथायमुपरिलिखितो ऽल्पो ऽङ्को घातमापकशब्देन
व्यवह्रियते ॥

––––––––––––––

बीजक्रिययया सम्पन्नानां फलानां सं-
ख्यामाचसाधारणत्वात् तान्येव व्यक्ते
सूत्रत्वेन परिणमन्ते   ॥ ५ ॥

––––––––––––––

उदाहरणे ।   यदा चतसृषु संख्यासु द्वितीया प्रथ-
मायां तावद्वारं वर्तते यावद्वारं चतुर्थी तृतीयायां तदा
ताश्चतस्रो ऽनुपातस्थाः स्युः ।   यथा (अ क ग घ)
एतासु चतसृषु (अ) अस्यां (क) इयं यावद्वारं वर्तते
तावद्वारं चेत् (ग) अस्यां (घ) इयं स्यात् तदैता अ-
नुपातस्थाः स्युः ।   कल्प्यतां तावत् (अ) अस्यां (क)

ख

द्वयं चिवारं ९ त्तदा कॢ=३ । अथच गॢ=३ । अथ यावन्तः पदार्थाः प्रत्येकं एकेनैव केनचित्समास्ते सर्वे मिथः समानादिति सिद्धमतः कॢ=गॢ ॥

अथच समयोः समगुणाने समतैवेत्यतिरोहितमतः (क+घ) अनेन पूर्वसमीकरणापचयोर्गुणितयोः सिद्धं (अघ=कग) । अनेनानुपातस्थेषु चतुर्षु राशिषु प्रथमचतुर्थयोर्वर्धो द्वितीयतृतीययोर्वर्धेन समानो भवतीति न्यायते । इदमेव महोपकारकस्य त्रैराशिकाख्यगणितस्य बीजम् । त्रैराशिके अनुपातस्थांस्त्रीन् राशीन् निर्दिश्य चतुर्थीं जिज्ञास्यते । तदा प्रश्नालापानुसारेण तान् यथास्थानं विन्यस्य द्वितीयतृतीययोर्वर्धः क्रियते । अथ स प्रथमचतुर्थयोर्वर्धेन तुल्य इति चतुर्थराशेरवगमाय स वर्धः प्रथमेन ह्रियते ॥

_____

कतिचन चेत्यमानविषया हि बीज-
गणितमन्तरेण नैव न्यायन्ते ॥ ६ ॥

_____

तद्यथा ।   यदङ्गुलमिता रेखा एकसंज्ञिका कल्प्येत
तर्हि अङ्गलद्वयमिता द्विसंज्ञिका स्यात् ।   अष्टाङ्गुलप-
रिमिता अष्टसंज्ञा स्यात् ।   अपिच या संख्या अका-
रेण चाप्यते तन्मिता रेखापि अकारेण द्योत्या स्यात् ।
तथा वर्गे चेचाणि घनचेचाणिच वर्णैर्न्नोपयितुं शक्य-
न्ते ।   ततो बीजगणितक्रियया   तद्वत्तिगुणा आ-
विर्भवन्ति ॥

आकारस्यानेके गुणा न कदाचिदपि अंकैर्न्नोपयितुं
शक्यन्ते ।   यथा तुल्यद्विबाहुकचिभुजस्य भूलम्बकोण-
योर्मिथः सम्बन्धः ।   अत्र रेखा सरला वक्राश्च भव-
न्ति ।   एतद्विषयिणी विद्या चेचमितिसंज्ञिका स्यात् ।
अथ सरलवक्ररेखालच्चणार्थं सूचम् ॥

या रेखा एकामेव दिशं गच्छेत् सा सरला
याच प्रतिपदं भिन्नदिशं गच्छेत् येन तच-

त्यौ यौ कौचनासन्नौ बिन्दू सरलरेखायां
न भवेतां सा वक्रा ॥ ७ ॥

अतः सरलवक्ररेखयोर्विन्द्वोरेकचस्थितयोस्तदव्यव-
हितोत्तरबिन्दू न कदाचिदेकच स्थातुं शक्नतः । यु-
क्लीदसंज्ञकस्य चेचमित्यां श्रीबापूदेवशास्त्रिणा संस्कृ-
तेन रचितायां टत्तपरिधिरूपिण्येव वक्रा व्यवहृता
नान्या ॥

युक्लीदानुमानमूलमिदं । ये चेचे मिथः संयुक्ते
सर्वतोभावेन परस्परं मिलतस्ते मिथः समाने इति ॥

यथा । प्रथमाध्यायस्य चतुर्थप्रतिज्ञेयम् । ययो-
स्त्रिभुजयोरेकस्य भुजद्वयतदन्तर्गतकोणौ क्रमेणापरस्य
भुजद्वयतदन्तर्गतकोणाभ्यां समानौ ते मिथः समाने
इति । अच त्रिभुजयोर्मिथः संयोजितयोस्ते सर्वतो-
भावेन मिथो मिलेतामित्यनुमीयते ॥

अच संशयः । यतः सरलवक्ररेखयोर्बिन्द्वोरेकच-

स्थितयोस्तदध्यवहितोत्तरविन्दू न कदाचिदेकत्र स्थातुं
शक्नुत इत्युक्तमतो यस्य चैतस्य मर्यादाः वक्राः
सन्ति यस्यच सरलास्तयोः परस्परं सर्वतोभावेन मे-
लनं न केनचिदपि विभागकल्पनेन सम्भवनीयम् ॥

भूगोलादिगणितसम्बन्धिविद्यानां वर्द्धनार्थं वक्ररेखा-
इतच्चैचार्थां सरलरेखाइतच्चैचैः साम्यं कर्तुं युक्लीदा-
ख्यस्य कालात् न्यूटनाख्यस्य कालपर्यन्तं द्विसहस्रवर्षेषु
अनेके उपकारकाः प्रकाराः कल्पिताः ॥

न्यूटनाख्यस्य प्रकारो लैब्नित्साख्यस्य जर्मणीदेशीय-
महाज्ञानिनो लेखनप्रकारेण परिष्कृतो अपरेषां लो-
पस्य कारणम्बभूव । स वैलक्षण्यपूरितगणितसञ्चकः
स्यात् ॥

स्तम्भोपरि स्थितो मयूरः स्तम्भतले स्थितं विषमा-
गच्छन्तं स्तम्भात्किञ्चिदन्तरे सर्पं दृष्ट्वा तन्ग्राहार्थं तदु-
पर्यपतत् । तत्र स्तम्भोच्चयं विषसर्पान्तरंच निर्दिष्टम् ।

सर्पमयूरयो: समकालगमनांशप्रमाणेच निर्दिष्टे । त-
थाच स मयूरो विलाल्कियत्यन्तरे सर्पं जग्राहेत्युदाजहार
भास्कराचार्यो लीलावत्यां । एतत्प्रश्नोत्तरं वैलच्चण-
गणितमन्तरेण न सिध्येत् । भास्करस्तु मयूरस्य याव-
त्सर्पसंयोगं सरलरेखायां गतिं मत्वा तदनुसारेण फल-
मानीतवान् । परं यदि मयूर: खसर्पसंयोगस्थानं
पूर्वं ज्ञातुं समर्थ: स्याच्चेत्तद् घटेत । मयूरो हि
प्रतिच्चणं खमार्गं विकरोति अतस्तज्ज्ञमनकक्कुप् प्रतिपदं
चलेत् । अत एव तज्ज्ञमनंचैकस्यां वक्ररेखायां भवेत् ।
तद्वक्ररेखानिर्णायकोपाया वैलच्चणगणितच्चेनैव ज्ञातुं
शक्यन्ते ॥

अथ वैलच्चणगणितविषये मध्यबीजज्ञज्ञेयमेकं सुग-
ममुदाहरणं कल्प्यते ॥

उच्चप्रदेशाङ्गुवमागत: पदार्थ: प्रथमेऽसुपादे पादो-
नैकादशहस्तमितदेशं अतिक्रामति । द्वयोरसुपादयो-

स्ततश्चतुर्गुणं । चिष्वसुपादेषु नवगुणमेवमग्रेऽपि ।
एवं पतत्पदार्थोतिक्रांतप्रदेशः कालवर्गानुसारं वर्द्धते ।
अनेनेदं चायते यदल्पतमकालेऽपि पतनवेगः समानो
न भवतीति अन्यथा बहुकाले बद्धन्तरं वेगस्य कथं स्या-
दिति । अथेष्टकाले दृष्टसमानगत्यातिक्रान्तं प्रदेशं नि-
श्चित्य सर्वदा तदनुसारेण गमनवेगं गणयन्ति परं यदि
पतत्पदार्थस्यासुतृतीयचरणान्ते संजातो वेगो चातथ्यः
स्यात्तदा सोऽवश्यं तत्पूर्वासुपादजातवेगादधिको भवेत्त-
दुत्तरासुपादजातवेगाद्यूनः स्यादिति । तदेतदर्थ-
मादौ न्यूटणाख्यकृतपदार्थविज्ञानशास्त्रस्याद्यप्रतिज्ञा उ-
पपाद्यते ॥

मिथः साम्यं यावच्चलतोर्ययोः प्रमा-
णयोर्भेदं उत्तरोत्तरं केनापि निर्हि-
ष्टमानेन निरन्तरं न्यूनो भवेत् ते अ-
न्ते मिथः समाने भवेतामिति ॥ ८ ॥

अचोपपत्ति: । यद्यन्ते ते मिथ: समाने न स्याता-
मित्युच्येत तर्हि तयोर्भेद: भ एतावान्स्यात् । अतो
न्यूनो नैव भवेत् परमेतत्पूर्वकल्पनाविरुद्धमतस्ते
अन्ते मिथ: समाने न भवेतामिति न किन्तु समाने एव
भवेतामित्युपपन्नम् ॥

अथ यदि पतत्पदार्थातिक्रान्तप्रदेशस्य पादोनैकाद-
शहस्तमितदेशो मापक: कल्प्येत तदा पतत्पदार्थेन (य)
असुपादैर्योवान् देशो ऽतिक्रम्यते स (यं) एतदनुसारे-
णास्तीति ज्ञायते ॥

अथ(च)अयमतिसूद्ष्मकालमानद्योतक: स्यात्| —ओ
अथच(ओ)स्थानात्प्रतित: पदार्थ:(य)असुपा- | —अ
दै:(अ) स्थानं प्राप्नुयात् (य+च) असुपादैश्च | —क
(क) स्थानं प्राप्नुयात (य+२च) कालेनच (ग) | —ग
स्थानं गच्छेदित्यादि कल्प्यतां तदा अतिक्रा- | —घ
न्तप्रदेशमानानि वक्ष्यमाणरीत्या ज्ञायन्ते । तथा-

हि । ओअ=य² । ओक=(य+च)² । ओग=
(य+२च)² इत्यादि । अतः

अक=ओक−ओअ =२यच + च²=(२य + च)च

कग =ओग − ओक =२यच+३च²=(२य+३च)च

गघ = ओघ − ओग =२यच+५ च²=(२+५च)च

इत्यादि ॥

अनेनेदं ज्ञायते (च) परिमितकालखण्डेषु क्रमेणाति-
क्रान्तप्रदेशाः २य+च २य+३च २य+५च इत्यादीना-
मनुसारेण भवन्तीति । अत्र बहुकाले वेगस्य बह्वन्तरं
स्यात् परं यथा यथा स्थानद्वयस्यान्तरमल्पं भवेत्तथा
तथा तच्त्यवेगयोर्भेदोऽपचीयेत ॥

अतो यथा यथा (च)कालस्य न्यूनत्वं स्यात्तथा तथा
प्रागुक्तातिक्रान्तप्रदेशप्रमाणानि अत्यासन्नानि भवेयुस्त-
त्प्रदेशान्ताश्चेष्टबिंदुसन्निहितास्युः पतनवेगाश्चाल्पान्तरा
भवेयुः ॥

अथात श्वान्ते यदा (च)स्य न्यूनता परमा स्यात् अ-

थीत् (च)स्य मार्जं शून्यं स्यात्तदा तानि प्रमाणानि अ-
पेक्षितप्रमाणेन समानानि भवेयुरिति पूर्वोक्तप्रतिज्ञया
सिध्यति ॥

अथ (च)स्य शून्यत्वे तानि सर्वाणि प्रमाणानि (२य)
भवेयुरतः (२य)द्वयं मर्यादा स्यात् । अनेन (य)असु-
पादान्ते पतत्पदार्थगम्यदेशः (२य) गुणितपादोनैकादश-
हस्तमितो भवेदिति ज्ञायते ॥

यथा । असुपादचयान्ते यदा य=३ तदाप्रभृति चे-
त्पतनवेगः समानो भवेत्तर्हि उत्तरासुचरणे स पदार्थः
सार्द्धचतुःषष्टिहस्तमितदेशं गच्छेत् । यतः २×३×$\frac{४३}{४}$=
६४$\frac{१}{२}$ । परं वस्तुतो वर्द्धमानगतित्वात्स पंचसप्तति
हस्तान् गच्छति ॥

अत्र (अोअ) रेखात्मको राशिः यः क्रमेण वर्द्धते ता-
दृशो ऽस्थिरराशिसंज्ञः स्यात् ॥

इहास्थिरराशयः (य) (र) (ल) इत्याद्यैर्वर्णैर्व्यज्यन्ते ।

अथच यैरस्थिरराशेर्मानं विलक्षणं भवति येषां चेष्ट-
सौक्ष्म्यस्यान्ते नास्ति ते (अक) (कग) (गध) इत्यादय:
खण्डा: वैलक्षणशब्दवाच्या:   ॥

अथ (य) स्य वैलक्षणं $_{d}$य अनेन द्योत्यते ।   एवमेव
र ल इत्यादीनां वैलक्षणानि क्रमेण $_{d}$र $_{d}$ल इत्यादिभि-
र्द्योत्यन्ते ॥

यस्य पूर्वोदितप्रश्नस्योत्तरं पूर्वोक्तरीत्या ज्ञातं तस्यो-
त्तरं वैलक्षणगणितेनैवं ज्ञायते ।   यदि (य) समानद्वद्ध्या
विकृतो भवेत्तदा (य)² अयं (२य) भवेदिति सिद्धं तदा-
चेत् (य) स्य वैलक्षणं $_{d}$य एतत्स्यात् तर्हि (य) स्य वैलक्षणं²
(२य$_{d}$य) भवेत्   ॥

अथ (२य $_{d}$य)² (य) अनयोर्मेलनेन वैलक्षणज्ञापको
ऽयं प्रथमो विधिरुत्पद्यते ।   यस्य वैलक्षणं ज्ञातव्यं तस्य
घातमापकं निरेकं कुर्यात् ततस्तस्मिन् घातमापकेन
तच्चत्यखतंचास्थिरराशेर्वैलक्षणेनच गुणिते ऽभीष्टं वै-
लक्षणं संपद्यते   ॥

अत्र कतिचिदुदाहरणानि प्रदर्श्यन्ते । d(य) अर्थात्
(य) अस्य वैलक्षण्यं (इयं dय) एतत्स्यात् । अथच d(य)=
(४य dय) । इत्यादि ॥

अत्र वैलक्षण्यस्थो यो गुणकः स उच्यते वैलक्षण्यप्रकृ-
तिरिति । यथा । (४यं dय) अत्र (४य) वैलक्षण्यप्रकृतिः
स्यात् ॥

यदा कस्यचिदस्थिरराशेर्मानमपराश्थिरराशिमाना-
धीनं स्यात्तदा स राशिरपरस्य कर्म स्यादित्युच्यते ॥

यथा । वर्गचेचस्य फलं तङ्कुजाधीनमतस्तङ्कुजस्य
कर्म भवेत् ॥

वैलक्षण्यगणितं नाम यस्य कस्यापि निर्दिष्टकर्मणो
वैलक्षण्यसाधनम् ॥

एतद्ध्रास्तं पूरितगणितसंज्ञं स्यात् । अर्थान्निर्दिष्टा-
द्वैलक्षण्यात्तत्कर्मण आनयनम् ॥

अत एव पूरितं पूर्वोक्तविधिविपर्यासेनोत्पद्यते ।

तथाहि । यस्य पूरितं ज्ञातव्यं तस्य घातमापकं सैकं कुर्यात् ततस्तस्मिन् सैकघातमापकेन तच्चत्यास्थिररा- श्येवैलच्चणेनच भक्ते ऽभीष्टं पूरितं संपद्यते ॥

अच चेत् (२य $_d$य) इदमुदाहियेत तदात: (य) इदं लभ्यते ॥

अथ ∫इदं पूरितचिह्नं भवति । अत: ∫२य $_d$य=य॒ ॥

या गणितरीतय: पूर्वं संच्चेपेण निरूपितास्ताभि- र्विविधकारणमेलनजनितं विरुद्धबलद्वयमियोजितपि- ण्डगत्यादिकमेकं कार्यं प्रायो निर्णेतुं शक्यते ॥

तच नानाकारणजनितैकाकार्यनिर्णयार्थं वक्ष्यमाण- नियमो ऽवश्यं ज्ञातव्य: । स यथा । यच कारणद्वयमे- लनादेकं कार्यं तच कारणद्वयं प्रत्येकं स्वातन्त्र्येण स्व- स्वप्रवृत्तिस्थलीयकार्यादन्यूनानतिरिक्तं कार्यं करो- तीति ॥

अयञ्च गतिविद्याया विषयेषु वस्तुत उपलभ्यते ।

२२

गतिविद्याविषयेष्वेककारणगता कार्यभक्तिरन्यकारणेन
न प्रतिबध्यते नवा नाश्यते । यथा । एक: पिण्डो
यावन्तं दिग्द्वयसम्बन्धिनं प्रदेशं यावता कालेन क्रमी-
त्पन्नबलद्वयप्रवर्तितो याति तावतैव कालेन तावन्तमेव
प्रदेशं विरुद्धदिग्द्वयगमनहेतुना युगपदुत्पन्नेनापि ब-
लेन प्रवर्तितो यातीति ॥

एवञ्च क्रमप्रवृत्तानेककारणोत्पन्नकार्याणि युगपत्प्र-
वृत्ततत्कारणोत्पन्नकार्येण समानि भवन्तीति नियमो
यथा गतिबलयोस्तथान्यचापि क्वचिदस्तीति ॥

परन्तु नायं नियम: सार्वत्रिक: यतो बहुधा द्र-
व्यद्वयसंयोगाज्जायमानं तृतीयं द्रव्यं खजनकद्रव्यद्वयगु-
णापेक्षया ऽत्यन्तविलक्षणगुणं भवति । तद्यथा । पारद-
धातुर्द्रवरूप: श्वेतवर्णो भवति गन्धकश्च पीतवर्णो भवति
अनयो: संयोगादत्यन्तरक्तवर्णो हिङ्गुल उत्पद्यते । एवं
रसरहित: सीसकाख्यो धातुरत्यम्लेन शुक्तेन संयुक्तो

मधुररसं विषविशेषं जनयति । उक्तनियमो यत्र ना-
स्ति तत्रत्या निर्जीववस्तुविषयिणी व्यवस्था रसायनशा-
स्त्रस्य विषयो भवति ॥

तत्रादौ रसायनशास्त्रीयव्यवहारसिद्धयेऽनात्मद्रव्या-
णि विभजते ॥

द्रव्याणि द्वेधा शुद्धमिश्रभेदात् ॥ ८ ॥

मिश्रद्रव्यं यथा पारदगन्धकात्मकं कपिश्रीर्षकम् ॥

अमिश्रद्रव्यं यथा गन्धको यस्य मिश्रितत्वविनिग-
मको न त्रायते ॥

मिश्रद्रव्याणि अमिश्रद्रव्यात्मकानि असंख्यानि ।
अमिश्रद्रव्याणि शताड्वाकिञ्चिददधिकानि । तेषां ता-
वड्भेदा उच्यन्ते ॥

अमिश्रद्रव्याणि द्वेधा धातुरूपा-
ण्यधातुरूपाणिचेति ॥ १० ॥

सुवर्णलोहपारदादिद्रव्याणि धातुरूपाणि । गन्ध-
कादीन्यधातुरूपाणि ॥

अधातुरूपाणिमिश्रद्रव्याणि द्विधा वा-
युरूपाण्यवायुरूपाणिचेति ॥ ११ ॥

वायुर्नाम स्पर्शवानसाधारणोष्णतां विनाप्यनियत-
देशव्यापनशक्तिमान् बद्धनियतसङ्कोचयोग्यश्च द्रव्य-
विशेषः । उष्णीकृतजलबाष्पमनियतदेशव्यापनश-
क्तिमद्भवति किन्त्वसाधारणोष्णत्याभावे तद्व्यापनशक्ति-
र्नश्यतीत्येतावान् तत्र भेदः ॥

वायवः साधारणादिभेदैर्बहुविधाः । तत्र यः प्रा-
णिनां श्वासोच्छ्वासोपयुक्तः स साधारणसंज्ञः । अन्ये
च वायुविशेषाः कथ्यन्ते ॥

अमिश्रवायुविशेषाश्च ख्या
रो भवन्ति ॥ १२ ॥

तेषां नामानि । प्राणप्रदः । जीवात्मकः । ज-
लकरः । हरिच्चेति ॥

प्रथमो वायुविशेषो ऽतः प्राणप्रद उच्यते यतो नहि
तं विना प्राणिनो जीवितुं शक्नुवन्ति । अस्मिन् न क-
श्चन खादः नापि गन्धः । अनेन पूर्णस्य काचपात्रस्य
वर्णः नहि साधारणवायुपूर्णकाचपात्रवर्णाद्भिन्नः । क-
स्तर्हि सामान्यवायोः प्राणप्रदस्यच भेद इति चेच्छ्रूय-
ताम् । दाह्यद्रव्याणि प्राणप्रदवायौ अतिशयेन ज्वल-
न्तीत्यस्ति तच भेदः ॥

अचिरादुपशमिष्यज्ज्वलने काष्ठखण्डे प्राणप्रदवायौ
धृते तत्काष्ठखण्डं अतिशयेन ज्वलितुं आरभते ॥

अत लोहो ऽपि शुष्केन्धनवत् ज्वलति ॥

प्रज्वलमानः प्रकाशदसंज्ञको ऽस्थ्यारम्भकावयववि-
शेषः प्राणप्रदवायौ प्रक्षिप्तः सूर्यवत्प्रकाशते नचैवं सा-
धारणवायौ ॥

२६

प्राणप्रदवायुं विना प्राणिनो न जीवन्तीत्युक्तं तदनु-
पपन्नं साधारणवायौ प्राणिजीवनदर्शनादिति चेन्न सा-
धारणवायौ प्राणप्रदवायोः सद्भावात् । नच साधा-
रणवायौ ज्वलनसामग्रीविशिष्टानां लोहादीनां दाहा-
पत्तिरुद्धेतोः प्राणप्रदवायोः सद्भावादिति वाच्यं तत्र-
त्यप्राणप्रदवायोर्जीवान्तकवायुना तदीयज्वलनशक्तुत्क-
र्षप्रतिबन्धकेन मिश्रितत्वादतिशयितदाहाभावोपपत्ते-
रिति ॥

एवं द्वितीयो जीवान्तकाख्यो वायुभेदो न गन्धवान्न
रसवान्नच काचपात्रसम्वृते ऽप्यस्मिन् रूपमुपलभ्यते
यथा प्रथमे । कस्तर्ह्यस्य प्रथमाद्विशेष इति चेत् शुद्धे
जीवान्तकवायौ प्रक्षिप्तं ज्वलद्द्रव्यमाशु निर्वाणमभवती-
त्येको विशेषः । अथ साधारणवायुर्यदि जीवान्तक-
वायुमात्ररूपः स्यात्तर्हि न किञ्चिदपि ज्वलेत् । यदिच
प्राणप्रदवायुमात्ररूपस्तद्येकस्फुलिङ्गस्पर्शात्सर्वं भस्म-
साद्भवेदिति मैवं साधारणवायोर्निर्णीतपरिमाणवि-

मेषानुसारेणोक्तवायुद्वयमिश्रिततया पूर्वोक्तदोषद्वया-
भावात् ॥

तच्च परिमाणां जीवान्तकवायोश्चत्वारोंऽशाः प्राण-
प्रदवायोरेकोंऽशः साधारणवायुघटको भवतीत्यत्र किं
मानमिति चेत् प्रत्यक्षपरीक्षासङ्कृतमनुमानमिति ब्रू-
मः । तद्यथा । जलपूर्णे पात्रे ज्वलन्ती वर्त्तिका तथा
स्थाप्या यथा जलोपरि तरेत् ततस्तदुपरि घटाकारं
काचपात्रपिधानं तथा स्थापनीयं यथा तस्याधरान्तो जले
किञ्चिद्विमज्जेत । एवं कृते अचिराज्ज्वलनोपकारो-
पच्चीषमग्नौ प्राणप्रदवायौ विकृते वर्त्तिका निर्वाति ज-
लञ्च तत्पात्रान्तःप्रदेशे बहुपञ्चमभागपर्यन्तमूर्ध्वं याति
तावति प्रदेशे प्राणप्रदवायोरभावञ्चानुमापयतीति ॥

तृतीयोऽपि वायुप्रभेद उक्तवायुद्वयवन्धरसरहितः
काचपात्रसम्भृततादृशायामनुपलभ्यमानरूपश्च । प्रा-
णप्रदवायुना सह तस्य संयोगाज्जलमुत्पद्यते तत्सूच-
नार्थं एतद्वायोः संज्ञा जलकरः स्यात् ॥

दाहविषये ऽयं प्राणप्रदवायोर्विपरीतधर्मो भवति । यदि ज्वलन्ती वर्त्तिका जलकरवायौ क्षिप्यते तदा त‍स्यामाशु निर्वाणायां जलकरवायुर्दीप्यते ॥

अयञ्च सर्वेष्वद्यावज्ज्ञातवस्तुषु मध्ये प्रकृष्टलघु‍र्भवति । अनेन पूर्णः पेशीविशेषः पश्चिगम्यप्रदेशादू‍र्ध्वं याति । एतद्वायूपष्टब्धेन महत्प्रमाणेन पेशीवि‍शेषेण कतिपयपुरुषसहितो मञ्चविशेषो बहूञ्चप्रदेशी‍परि नेतुं शक्यते ॥

एवं चतुर्थो वायुभेदो चरिताख्य उक्तेभ्यः शुद्धवायु‍विशेषेभ्यो बहुभिर्विर्विशेषैर्विशिष्यते । अस्य वर्णो ह‍रित्प्रायो रसगन्धौचात्यन्तोत्कटौ दुःखप्रदौच ॥

वक्ष्यमाणाः कतिचिद्द्रव्यविशेषा एतद्वायुसंयोगमाचेष्ट ज्वलन्ति । ते यथा । ताम्रस्यातिचिपिटपर्च प्रकाश‍दसंज्ञकश्च पारदादिचेति ॥

अयं प्राणिनां जीवननाशको द्वचजातीनां शौक्लद्वहे‍तुश्च । अत एव वस्त्रादिशुक्लीकरणे परमोपयोगी ॥

किञ्च रोगिशालाकारागारादिष्वयं वायुर्महत उप-
काराय भवति तदत्यानां रोगवशात् कारणान्तराद्वा
पूतिगन्धादिनाशकत्वात् ॥

धातुवाव्यंतिरिक्तानि शुद्धद्र-
व्याणि नव भवन्ति ॥ १३ ॥

एतानि तेषां नामानि । यथा । अङ्गार: ।
टङ्कोपादानं । अग्निप्रस्तरजनक: । गंधक: । चा-
न्द्र: । प्रकाशद: । पूत: । अरुण: । काचप्रज-
नक: ॥

ध्मातकाष्ठस्य दहने जलेन निर्वाणे ऽवशिष्टोऽङ्गारो
भवति श्यामवर्ण: ॥

टङ्कावयव: कपिशचूर्णविशेष: टङ्कोपादानम् ॥

लौहखण्डाघाताद्ग्निजनकस्य प्रस्तरविशेषस्यावय-
वभूत: कपिशवर्णश्चूर्णविशेष: अग्निप्रस्तरजनक: ॥

दाह्यः पीतवर्णः सर्वैर्ज्ञातो द्रव्यविशेषः गन्धकः ॥

बहुभिर्धर्मैर्गन्धकतुल्यो ऽतिदुर्मिलो द्रव्यविशेष-
स्वान्द्रः ॥

सिक्थकसट्टशरूपः तमसि प्रकाशः क्वथितजलोष्ण-
तान्यूनोष्णसंयोगेन घर्षणादिना वा दाह्यः घर्षणज्चलनी-
यदीपशलाकानिर्माणोपयुक्तो द्रव्यविशेषः प्रकाशदः ॥

कृष्णारुणोऽतिदुर्गन्ध्ययुक्तो द्रवद्रव्यविशेषः पूतः ॥

नैसर्गिकभास्वरकृष्णवर्णोऽग्निसंयोगेन बाष्पत्वदशा-
यामरुणवर्णो द्रव्यविशेषोऽरुणः ॥

काचनाशकद्रवद्रव्यावयवत्वेन कल्पितो वस्तुविशेषः
काचप्रजनकः ॥

धातवष्षट्चत्वारिंशत ॥ १४ ॥

तच्च सुवर्णरजतताम्रसीसकाद्यो धातवः कतिचित्
सम्यग्विज्ञाताः कतिचिच्चातिदुर्मिला अनावश्यकनिरू-

पषाश्च । अतो येषां खरूपज्ञानमिहावश्यकां त एव धातवो ऽत्र निरूप्यन्ते ॥

सुवर्णाधिकगुरूरजतवर्णः गुरुतमाख्यो धातुः ॥

पारदाख्यो धातुस्साधारणोष्ण्यसंयोगे जलवद्द्रवरूपो ऽत्यन्तोष्ण्यसंयोगे बाष्परूपोऽतिशीतदेशे तथा घनो यथास्य पचाणि कर्तुं शक्यन्ते ॥

रङ्गं रजतवर्णं सीसकवन्मृदुच ॥

त्रस्ताख्यो धातुः सीसकवर्णोऽतितप्तादशायां साधारणवायौ दग्धः तूलवद्वायुना सर्वत्र छिप्यते ॥

लघुतमाख्यो धातुरत्यन्तलघुत्वात् जले तरति । अयंच जले वर्त्तमानः खयमेव ज्वलति ॥

लवणकराख्यो धातुः जले चिप्मस्तरति परन्तु न लघुतमवज्ज्वलति ॥

शर्कराकराख्यो धातुर्यस्साच्चूर्णमुत्पद्यते ॥

अथ प्राणप्रदादिभिः शुद्धद्रव्यैर्भूमण्डलोपलब्धानि सर्वाणि निर्मितानि सन्ति ॥

यद्द्रव्यं यद्द्रव्यजन्यं तद्द्रव्यस्य तत्तद्द्रव्यजन्यताबोधक-
पदेन व्यवहारोऽत्यन्तोपकारक इत्यत आह ॥

रसायनशास्त्रे मिश्रद्रव्याणि स्व-
जनकतत्तद्द्र्योत्पन्नलव्यञ्जकसं-
ज्ञया व्यवहर्त्तव्याणि ॥ १५ ॥

उदाहरणे । हरितवायुना सह लवणाकराख्यधा-
तुमेलनात् साधारणलवणमुत्पद्यत इत्यत्र शास्त्रे सा-
धारणलवणस्य संज्ञा लवणाकरहरितज इति । अ-
र्थात् । लवणाकरस्य हरितवायोश्च मेलनात् लवण-
स्योत्पत्तिरिति ॥

अथ यथोद्देशक्रमं प्राणप्रदवायुमारभ्य अमिश्रद्र-
व्याणां परस्परमेलनकार्याण्युच्यन्ते ॥

साधारणवायौ प्राणप्रदवाय्वंशेन सह जीवान्तका-
ख्यवायोश्चत्वारोऽश्च भवन्तीत्युक्तं । अत्र साधारण-
वायुस्वरूपं कार्यं रासायनिकं नोच्यते यतोहि तयो-

मेलने ऽपि विशेषो नोत्पद्यते । यच्चतु प्रत्येकज्ञात-
विशेषयोः पदार्थयोर्मेलनादुत्पन्नं कार्यं नानुमानेन
ज्ञातुं शक्यते तच्चैव रासायनिकसम्बन्धो भवतीति पूर्वं
सूचितम् ॥

जीवान्तकवायुना सह प्राणप्रदस्य मेलनाच्च तद्वायु-
परिमाणभेदानुसारेण पञ्च कार्यविशेषा उत्पद्यन्ते ॥
तेषुचोत्तरोत्तरं एकांशेन प्राणप्रदवाय्वाधिकं ज्ञा-
तव्यम् । अतस्तेषां वर्णनं क्रियते । तेष्वाद्यो यथा ॥
स जीवान्तकप्राणप्रदयोर्योगो यमुच्छासश्वासक्रियाया
विषयीकुर्वन् पुरुषो ऽत्यन्तं हसति कतिपयच्चणपर्यन्त-
मुन्मत्त इवच भवति ॥

द्वितीयो यथा ॥

स उक्तवाय्वोर्योगो यो नीरूप: साधारणवायुसंयो-
गदशायां प्राणप्रदवायोरधिकं भागमाकर्षति रक्तवर्ण-
श्च भवति । अचैवाङ्गार: प्रकाशदश्च सबहुज्वालं ज्वल-
ति अयंतु न श्वासोच्छ्वासक्रियाविषयो यत उरोवर्तिना

ङ

साधारणबायुना संयोगकाले पूर्वोक्तं रक्तवर्णं विषखभावं वायुविशेषं जनयति ॥

तृतीयो यथा ॥

स उक्तवाय्वोर्योगविशेषो यो महता प्रयत्नेन बिना नार्जयितुं शक्यः अस्मिन्प्रकरणे विशेषतस्तस्य निरूपणं नावश्यकम् ॥

चतुर्थो यथा ॥

पूर्वोक्तो रक्तवर्ण उक्तवाय्वोर्योग: यो जलेन संयुक्त- स्तश्चैव लीयते ॥

पञ्चमो यथा ॥

स उक्तवायुद्वययोगो यो जलमिश्रितात्यन्ताम्लरस- विशेषस्वरूपो मांसं दहति यस्मिंश्च लोहादिकमपि लीयते ॥

अथ यद्द्रव्यं यद्द्रव्यजन्यं तद्द्रव्यस्य तद्द्रव्यजन्यताबोधक- पदेन व्यवहारो ऽस्मिन् शास्त्रे कर्त्तव्य इति पूर्वमुक्तं तथा उक्तवायुद्वयमिश्रणजनितानां परस्परमत्यन्तविलक्षणा-

नां द्रव्याणां व्यवहारो मिश्रणावयवानामाधिकन्यून-
त्वज्ञापकपदेनच कर्त्तव्यः । यथा । द्वितीययोगः प्रथ-
मयोगमपेक्ष्याधिकेन प्राणप्रदवायुना विशिष्ट इतिचेत्तो-
स प्राणप्रदवाय्वाधिक्यबोधकपदेन व्यवहर्त्तव्य इति ॥

अथ नानाविधवस्तुसंयोगसम्बन्धिनां विविधधर्मो-
णां निरूपणापूर्वकं उक्तविधसंकेतरचनोपायानां नि-
रूपणं क्रियते ॥

एतावत्परिमाणेनामुकेनैतावत्परिमाणम-
मुकं संयुक्तं भविष्यति न न्यूनाधिकमिति
नियमः केषुचिद्द्रव्येषु न दृष्यते ॥ १६ ॥

यथा । एकपलमितं दुग्धं एककर्षमितेन जलेन
संयुक्तं भवति एकपलमितं जलंच एककर्षमितेन दुग्धेन
संयुक्तं भवति नचाच कश्चिन्नियमो ऽस्तीति ॥

क्वचित्तु न्यूनतायामनियम आ-
धिक्ये त्ववधिर्भवति ॥ १७ ॥

यथा । अमुकपरिमाणे जले ऽमुकपरिमाणां लवणं लीयते इत्येतस्मिन्विषये लवणस्याधिक्येऽवधिरस्ति नतु न्यूनतायाम् ॥

उक्तं वस्तुसंयोगस्य प्रकारद्वयं न विशेषतो वस्तुतो रसायनशास्त्रस्य विषयो यत: तच्च संयुक्तद्रव्यद्वयस्य असं‐ योगावस्थायां दृश्यमाना गुणा उत्तरकाले ऽनुवर्त्तन्ते ऽथवा न सर्वथा तेषां अन्यथाभावो भवति ॥

रसायनशास्त्रीयविधिनुसारेण संयुज्यमान‐ योरेकजातीयकार्यौत्पादकयो: पदार्थयो: संयोगो नियतपरिमाणयोरेव भवति यच्च पुन:संयुक्ताभ्यां पदार्थाभ्यां नानाजातीयानि मिश्रणफलान्युत्पद्यन्ते तच्च द्वितीयादिमि‐ श्रणे एकद्रव्यस्य प्रथमसंयोगस्थलीयमेव प‐ रिमाणं अपरस्यतु प्रथमसंयोगस्थलीयपरि‐ माणापवर्त्यमेव परिमाणं भवतीति ॥१८॥

यथा । जीवान्तकवायुना सह प्राणप्रदस्य मेलनात्
पञ्च मिश्रपदार्था उत्पद्यन्त इत्युक्तम् । तेषु प्रत्येकं
जीवान्तकश्चतुर्दर्शमितो भवति प्राणप्रदश्च क्रमेण । ८ ।
१६ । २४ । ३२ । ४० । एतत्परिमाणो भवति ॥

यत्परिमाणकौ पदार्थौ प्रत्येकं जल-
करवायुना संयुज्येते तत्परिमाण-
कावेव तौ मिथः संयुज्येते ॥ १८ ॥

यथा । जलकरस्यैकेनांशेन हरिद्रायाः पञ्चविंशद्भा-
गाः संयुज्यन्ते प्राणप्रदवायोश्चाष्टौ । तदा हरिद्रायाः
पञ्चविंशद्भागाः प्राणप्रदस्य । ८ । ३२ । ४० । ५६ ।
एभिर्भागैः पृथक् संयुज्यन्ते । अथ । ८ । ३२ । ४० ।
५६ । एते ऽङ्का अष्टानामव्यवहितापवर्त्तनानि न भव-
न्ति । अव्यवहितापवर्त्तनाङ्कास्तु । ८ । १६ । २४ ।

। ३२ । ४० । ४८ । ५६ । एते भवन्ति । अत:
। १६ । २४ । ४८ । एते प्राणप्रदस्य भागा हरि—
द्वायुना संयुक्ता: केषुचित्पदार्थेषु वर्त्तेरन् ये पदार्था
अस्माभिर्द्यावधि न ज्ञायन्ते ॥

रसायनविधिना जायमान: संयोगो नियतपरिमाण-
योरेव भवतीति नियमदर्शनात् भौतिकपदार्थानां
सर्वेऽवयवा दिगादिवद्ध विभागयोग्या: अपितु अत्यन्त-
सूक्ष्मा अशक्यविभागाश्च केचिदवयवास्सन्तीति कल्प्यते
तेच परमाणव उच्यन्ते ॥

किञ्च परमाणूनामतिसूक्ष्मत्वेऽपि तेषां सर्वथा मह-
त्त्वराहित्यं न कल्प्यते तच्च द्वितीयाध्याये चत्वारिंशत्त-
मे सूत्रे विस्तरेण निरूपितम् ॥

अपिच यद्यपि सुवर्णादीनामेकजातीया: परमाणबो
गुरुत्वे महत्त्वेच परस्परं तुल्यरूपा: तथापि ते भिन्नजा-

तीयलोहादिपरमाणुभ्यो गुरुत्वे विलक्षणा एव भव-
न्तीति ॥

यदा रसायनविधिना संयुक्ताभ्यां पदार्थाभ्यामेकजा-
तीयमेव कार्यमुत्पद्यते तदा तयोरेकस्यैकैकः परमाणु-
रपरस्यैकैकपरमाणुना संयुज्यते यच्च पदार्थयोर्मेलना-
द्बहुजातीयकार्योत्पत्तिस्तत्रैकः परमाणुरेकेन द्वाभ्यां
त्रिभिश्चतुर्भिर्वा संयुक्तो भवति ॥

संयुज्यमानपदार्थपरिमाणस्य संख्यया तदीयपरमाणू-
नां परस्परापेच्चया गुरुलघुभावोऽनुमीयते । तद्यथा ।
एकमेकपलमितमपरच्चाष्टपलमितमिति द्वे संयुक्ते यच्चैकं
नवपलमितं कार्यं जनयतो यच्च द्वयोः परमाणवः प्र-
त्येकमेकैकेन संयुज्यन्ते तत्रैकपलमितद्रव्यस्य परमाणु-
रष्टपलमितद्रव्यपरमाणोरपेच्चयाष्टगुणितगुरुरिति स्प-
ष्टम् ॥

मुख्यद्रव्यपरमाणूनां परस्परापेच्चगुरुलघुभावस्य स्प-
ष्टीकरणार्थं तदङ्कैःसह तन्नामानि प्रदर्श्यन्ते ॥

प्राणप्रद: ८	प्रा	ताम्रम् ३२	ता
जीवान्तक: १४	जी	लोहम् २८	लो
जलकर: १	ज	सीसकम् १०४	सी
हरित: ३५	ह	गुरुतमम् ८८	गु
अङ्गार: ६	अ	पारद: १००	पा
टङ्कोपादानम् ११	ट	दस्तम् ३२	द
गन्धक: १६	ग	लघुतमम् ४०	लघु
प्रकाशद: ३२	प्र	लवणाकरम् २३	ल
सुवर्णम् ८८	सु	शर्कराकर: २०	श
रजतम् ११०	र		

बीजगणित इव रसायनकार्यविचारेऽपि तन्तद्द्रव्याणां सङ्क्षिप्तसङ्केतकरणमुपकारि यथा पूर्वसूच्यां प्रदर्शि-तम् ॥

तेच सङ्केता: पदार्थद्वयमेलनजनितकार्यगुरुत्वज-नकप्रत्येकपदार्थगुरुत्वस्य स्वसन्निहितलिखितसंख्यायोगं

बोधयन्ति यथा प्रा द्वयं प्राणप्रदस्याष्टसंख्येषु भागेषु
सङ्केतितो जी इति जीवान्तकस्य चतुर्दंशसु ज इतिश्च
जलकरस्यैकस्मिन्नित्यादि ॥

एते एव सङ्केता मिलितास्न्तो मिश्रणफलस्य सङ्केता
भवन्ति । तद्यथा । ल इति लवणकरस्य चयोविंशतिषु
भागेषु च इतिच हरितस्य पञ्चविंशत्सु भागेषु सङ्केतितम्
तद्वयमपि ल + च इतिरूपेण मिलितं सत् अष्टप-
ञ्चाशद्भागान् लवणस्य बोधयति । इयं संख्या लवणपर-
माणुगुरुत्वस्य संख्यां बोधयति । एवं सर्वच मिश्रण-
फले परमाणुगुरुत्वसंख्यावगमप्रकारो बोध्यः ॥

अथैवमुक्तप्रकारेण मिश्रणफलसम्बन्धिनः सिद्धान्ता-
नभिधायेदानीं प्राणप्रदजीवान्तकवायुद्वयमिश्रणफला-
नि निर्णीयन्ते ॥

१	जो + प्रा	२२
२	जो + २प्रा	३०
३	जो + ३प्रा	३८
४	जी + ४प्रा	४६
५	जी + ५प्रा	५४

एषु पञ्चस्वपि मिश्रणफलेषूक्तसङ्केतेन घटकद्रव्य-
परमाणुपरिमाणं ज्ञाप्यते । घटकद्रव्यपरिमाणतज्ञ-
न्यत्वादिबोधकेनच सङ्केतेन रसायनशास्त्रे व्यवहार
दृष्टः । अथेदमधिकं विचार्यते । तथाहि ॥

उक्तेषु प्राणप्रदजीवान्तकयोर्मिश्रणफलेषु कानिचि-
दम्लरूपाणि कानिचित्तु न तथा । अथायमवश्य-
निरूपणीयो भेदः स्वबोधिकां संज्ञामपेचत इत्याह ॥

अम्लानम्लरूपताबोधकैस्सङ्केतैर्मिश्र-
णफलानां व्यवहारः कर्त्तव्यः ॥ १८ ॥

जीवान्तकप्राणप्रदमिश्रणफलेषु त्रीणि अनम्लरू-
पाणि अतस्तेषूत्तरोत्तरं केवलं प्राणप्रदपरिमाणाधि-
क्यबोधकसङ्केतेन व्यवहारः क्रियते यथा ॥

जीवान्तकस्य	प्रथमप्राणप्रदानम्लविकारः	
जीवान्तकस्य	द्वितीयप्राणप्रदानम्लविकारः	
जीवान्तकस्य	परमप्राणप्रदानम्लविकारः	

उक्तवायुदयमिश्रणफलेषु द्वे अम्लरूपे । तयोः सङ्केतायेदमारभ्यते । तथाहि । प्राणप्रदवायुना संयुक्तं द्रव्यं यदैकजातीयमम्लमारभते तदा सोऽम्लः ठन्प्रत्ययान्तेन पदेन व्यवहर्त्तव्यः । यदाच उक्तद्रव्येण द्विजातीयौ अम्लौ उत्पाद्येते तदा तयोर्मध्येऽधिकप्राणप्रदविशिष्टो ऽम्लः ठन्प्रत्ययान्तेन न्यूनप्राणप्रदविशिष्टो ऽम्लश्च यप्रत्यान्तेन व्यवहर्त्तव्य इति तृतीयपञ्चमौ च पूर्वोक्तवायुमिश्रणफलभेदावम्लरूपौ क्रमेण न्यूनाधिकप्राणप्रदविशिष्टौचातस्तयोर्जीवान्तक्यो जीवान्तकिकश्चेति संज्ञाद्वयं क्रमेण स्यात् । एवंच पञ्चानामपि वायुमिश्रणफलानामिमाः संज्ञाः ॥

तथाहि ॥

। १ ।	जीवान्तकस्य प्रथमप्राणप्रदानम्लविकारः ।
। २ ।	जीवान्तकस्य द्वितीयप्राणप्रदानम्लविकारः ।
। ३ ।	जीवान्तक्योऽम्लः ।
। ४ ।	जीवान्तकस्य परमप्राणप्रदानम्लविकारः ।
। ५ ।	जीवान्तकिको ऽम्ल इति ॥

एवं जीवान्तकप्राणप्रदनामकस्य प्रथमतत्त्वद्वयस्य प-
रस्परमिश्रणफलानि निरूप्य इदानीं जलकरसंज्ञकेन
तृतीयेन तत्त्वेन सह उक्ततत्त्वद्वयस्य प्रत्येकं मिश्रणफ-
लानि निरूप्यन्ते ॥

जलकरः प्राणप्रदश्च प्रतिपरमाणु संयुक्तौ जलं जन-
यतः । अस्योपपत्तिर्वक्ष्यते । अस्य मिश्रणफलस्य
सङ्केतो ज + प्रा इति । पूर्ववत् जलपरमाणुगुरुत्वं च
नवसंख्याकं ज्ञेयम् ॥

जलकरौ जीवान्तकेन संयुक्तो नवसागराख्यजनकं
उग्रगन्धं वायुविशेषं जनयति यः स्वभावतो ऽनुरागप्रे-
रित इव जलेन सह संयुज्य लीयते । इह जलकरस्य
द्वयः परमाणवः एकेन जीवान्तकवायुपरमाणुना सह
संयुज्यन्ते अत उक्तवायोस्सङ्केतो ३ ज + जी इति
तदीयपरमाणुगुरुत्वं च । १७ । सप्तदशसंख्याकं
बोध्यम् ॥

हरितवायुनामकं चतुर्थं तत्त्वं प्राणप्रदजीवान्तकज-
लकरवायुभिः संयुज्यमानं विविधानि मिश्रणफलानि
जनयति । तत्र हरितजलकरवाय्वोः संयोगफलमिदा-
नीं निरूप्यते तत्त्वोग्रगन्धं वायुस्वरूपं जलकरहरितकं
अम्लं अनुरागप्रेरितमिव स्वभावतो जलेन संयुज्य ली-
यते । अस्य संक्षिप्तसंकेतश्च ज + ह इति एतद्दी-
यपरमाणु गुरुत्वंच षट्त्रिंशत्संख्याकं बोध्यम् ॥

पूर्व एकादशसूत्रे अधातुरूपाण्यमिश्रद्रव्याणि वाय्व-
वायुभेदेन द्वेधा विभक्तानि । तत्र चतुर्णा अमिश्रवायु
भेदानां प्रत्येकं परस्परमिश्रणफलानि निरूपितानि ।
इदानीं क्रमप्राप्तानामवायुरूपाणाममिश्रद्रव्याणां मि-
श्रणफलानि निरूप्यन्ते ॥

तत्र प्रथमोऽङ्गार: । अयं प्राणप्रदवायुना संयुक्तो-
ऽङ्गारिकं अम्लं प्राणिप्राणहरं वायुविशेषमुत्पादयति ।
अयंच गुरुत्वाज्जलवत्पाचात्पाचान्तरे निर्गमयितुं श-

क्यते । अत एव यच्चायं क्वचिन्निरिगुह्वाप्रदेशे स्वभावतो
वर्त्तते तस्याधः प्रदेशे एव वर्त्तते नतूर्ध्वेभागे तच्च स्वा-
दिः प्राणी अनच्चत्वान्म्रियते मनुष्यादिजातिस्तूच्चत्वान्न
प्राणं त्यजति । अस्याङ्गारिकाम्लस्य सच्चिम्रसंकेतो अं
+ प्रा इति एतदीयपरमाणुगुरूत्वं च द्वाचिंशत् सं-
ख्याकं भवति प्राणप्रदवायौ ज्वलिताङ्गीरादाङ्गारिका-
म्लोत्पत्तेर्हीरकोऽङ्गारारब्ध इति निश्चीयते ॥

अङ्गारो जीवान्तकवायुसंयुक्तो ज्वलनशक्तिमन्तं वि-
चित्रगंधं जीवान्तकाङ्गारजाख्यं वायुविशेषं जनयति ॥

अङ्गारो जलकरसंयुक्तो ज्वलनशक्तिमंतं जलकरा-
ङ्गारजाख्यं वायुविशेषं जनयति । अयंच क्वचित्तडाग-
स्थपंकोत्सारणादुपलभ्यते । क्वचिच्च पंचनद्ददेशीयज्चा-
लामुखीप्रदेशादौ अयमेव भूविवरात्सदा ज्वलनमहा-
परिमाणो निर्याति । एतज्जातीय एव वायुविशेषः
प्रस्तराङ्गारेभ्यो निष्कृष्टुं शक्यते एषु दिवसेषु यूरोपप्र-
देशे रात्रौ रथ्याप्रकाशनं तया प्रयुज्यतेचेति ॥

गंधकं प्राणप्रदवायोस्त्रिभिः परमाणुभिः संहृष्टं गां-
धकिकमम्लं वायुरूपं जले लययोग्यं उत्पादयति ।
तस्य संकेतो गं + प्रा इति परिमाणुपरिमाणञ्च च-
त्वारिंशत्संख्याकं बोध्यमिति ॥

प्रकाशदः प्राणप्रदस्य पञ्चभिः परमाणुभिः संयोगे
सति प्राकाशदिकमम्लं जनयति । तस्य सङ्केतः प्र +
५प्रा इति अतस्तदीयपरमाणुगुरुत्वं द्विसप्ततिसंख्याकं
ज्ञेयम् ॥

प्रकाशदो जलकरसंयुक्तः साधारणवायौ स्थितो ज्व-
लनस्वभा ॱ वायुविशेषं जनयति । अयञ्च प्रकाश-
दस्य जलकरज इति व्यवह्रियते ॥

दशमसूचे अमिश्रद्रव्याणि धात्वधातुभेदेन द्वेधा वि-
भक्तानि । तच्च वाय्ववायुभेदेन द्विविधानामधातुद्र-
व्याणां मिश्रणफलानि निरूपितानि । इदानीं धातूनां
स्वभेदैः सह उक्तद्रव्यैश्च सह मिश्रणफलानि निरूप्यन्ते ॥

लोहादिनानाधातूनां प्राणप्रदेन सह मिश्रणफलं

किट्टमिति लोके व्यवह्रियते । सुवर्णं रजतञ्च प्राण-
प्रदेन संयोगं नेच्छतीव अत एव सुवर्णरजते व्यावहा-
रिकमुद्रादिरचनाय सम्यगुपयुज्येते ॥

लोकप्रसिद्धरक्तिम लोहकिट्टं द्वाभ्यां लोहपरमाणुभ्यां
चिभिश्च प्राणप्रदपरमाणुभिरारब्धं भवति । यदि द्व-
योस्त्वयन्तुद्र्वोकस्य सार्द्धमिति प्रमाणानुसारेण उक्तलोह-
किट्टं लोहस्य सार्द्धप्राणप्रदजमित्युच्यते अस्य सङ्केतो
२ लो + ३ प्रा इति । अत एव तत्परमाणुगुरुत्वं
चत्वारिंशन्मितंभवतीति । अथ प्राणप्रदलोहयोर-
न्यदेकं मिश्रणफलं लोहस्य प्रथमप्राणप्रदजमितिसंज्ञ-
कमस्ति यचैको लोहपरमाणुरेकेन प्राणप्रदपरमाणुना
संयुज्यते अतस्तस्य सङ्केतो लो + प्रा इति अतश्चास्य
परमाणुगुरुत्वं षट्त्रिंशन्मितं बोध्यम् । इदञ्च त-
त्सायोगोलाभिघातोत्पतितश्यामलोहमलांशस्वरूपम्भ-
वति ॥

सीसकस्य प्रथमप्राणप्रदजं पीतं भवति द्वितीयप्राण-
प्रदजं च रक्तं सिंदूरं इति लोके प्रसिद्धं भवतीति ॥

पारदस्य परमप्राणप्रदजं रक्तवर्णं भवति । पारद-
हरितवाय्वो: संयोगाद्वे मिश्रणफले भवत: तच्च प्रथम-
हरितपारदज: श्वेतवर्णश्चूर्णविशेषो महौषधं परम-
हरितपारदजं रसकर्पूर इतिलोकप्रसिद्धं महाविषं ॥

पारदधातुर्द्वैरूप: श्वेतवर्णो गन्धकसंयोगादत्यन्त-
रक्तवर्णं हिङ्गुलं जनयतीति अष्टमसूच्याख्यावसाने
उक्तं । अत्र शास्त्रे तत् पारदस्य द्वितीयगन्धकजमिति
संज्ञकम्भवति ॥

पारदो ऽनेकैर्धातुभि: सह शीतावस्थायामपि संयु-
ज्यते । रजतपाचार्णां सुवर्णलेपसिद्धये सुवर्णमिश्रित:
पारद उपकरोति ॥

लघुतमं प्राणप्रदसंयोगेन शुक्लवर्णं अतितीच्ष्णरसं
द्रव्यविशेषं जनयति इदंच मांसं कर्तयति अस्य संज्ञा
लघुतमा इति भवति ॥

लवणाकरप्राणप्रदसंयोगफलं लघुतमप्राणप्रदसंयोगे-
नात्यन्तसदृशं अस्ति । तस्य सङ्केतो ल+प्रा इति अतो-
ऽस्य परमाणुगुरुत्वं एकविंशन्मितं भवति ॥

हरितवायुना सह लवणाकरमेलनात्साधारणलवण-
मुत्पद्यत इति पूर्वमुक्तं पञ्चदशसूचव्याख्याने । अत्र शास्त्रे
तस्य नाम लवणाकरस्य हरितज इति । अस्य सङ्केतो
ल+ह इति अतोऽस्य परमाणुगुरुत्वं अष्टपञ्चाशन्मितं
भवति ॥

शर्कराकरप्राणप्रदसंयोगात् लोकप्रसिद्धाः शर्कराः
उत्पद्यन्ते । तस्य सङ्केतः श+प्रा इति अतोऽस्य परमाणु-
गुरुत्वं अष्टाविंशतिमितमस्ति ॥

एतावता द्रव्यद्वयारब्धमिश्रणफलानि निर्णीय तद्-
धिकद्रव्यमिश्रणफलानि अधुना निरूप्यन्ते । विंशति-
मसूचे उम्लानम्लरूपविभागद्वयस्य तदीयाम्लवाऽन-
म्लत्वबोधकसंज्ञया व्यवहारः कर्त्तव्य इत्युक्तम् । अम्ल
इति संज्ञयाऽम्लानामम्लरसवच्चरूप एको विशेषधर्मो
बोध्यते अन्यश्चाम्लानां विशेषधर्मोऽयं भवति यत्तेषां

चुम्बजपादिपुष्परससंयोगजनितनीलिम्ना जलेन संयोगे
तज्जलस्य रक्तताऽऽपादनमिति कानिचिदनम्लानि द्रव्य-
द्वितयमात्रसंयोगफलानि लघुतमावदतितीच्छारसानि
सन्ति तानिचोक्तविधजलस्य खसंयोगेन हरिततां जन-
यन्ति । एवंविधैरनम्लविकारैरन्यैश्च धातुप्राणप्रदसं-
योगफलै: किट्टैस्संयुक्तान्यम्लानि बहुविधमिश्रणफल-
जनकानि भवन्ति तन्निरूपणाय सूत्रम् ॥

अम्लानां धातुप्राणप्रदसंयोगफलै: किट्टै:
संयोगाल्लवणविशेषा जायन्ते ॥ २१ ॥

खाद्यलवणसाट्ट्याख्याल्लवणशब्दो ऽत्र लाक्षणिक: लव-
णविशेषा इति न सर्वा लवणजातिरम्लधातुप्राणप्रदसं-
योगफलभूतकिट्टृमिश्रणजन्या अपितु कतिपयलवणजा-
तिरेव यत: खाद्यलवणं हरितलवणकरमिश्रणजन्यं
भवति इदञ्च पूर्वमुक्तं । अथैते लवणविशेषा उच्यन्ते ॥
केचिदम्ला इग्लन्तेन केचिच्च यान्तेन पदेन व्यवहर्तव्या
इत्युक्तं अष्टाविंशतितमसूत्रव्याख्याने । तच्चेग्लन्तपदव्य-

वह्नताम्ललजनितलवणजातिर्यिदन्तेन यान्तपदव्यवहृता-
म्ललजनितलवणजातिश्वेदन्तेन पदेन व्यवहर्त्तव्या ।
यथा । जीवान्तर्किकाम्ललजं लवणं जीवान्तकायितमिति
यथाच जीवान्तक्याम्ललजं लवणं जीवान्तकितमिति ॥

अम्ललानि किट्टैरेव संयुज्यन्ते न धातुमाचेषेतिनिय-
मदर्शनादनेकजातीयकिट्टाम्ललमिश्रणजनितातिरिक्तानां
लवणानां किट्टजन्यताबोधकपदेन व्यवहारोऽनावश्यक:
अतश्च रजतप्राणप्रदजस्य जीवान्तकायितमितिसंज्ञापे-
च्चया लाघवाद्रजतस्य जीवान्तकायितमित्येव संज्ञा कर्त्त-
व्या । परन्तु लोहप्रथमप्राणप्रदजस्य गन्धकायिते
लोहसार्द्धप्राणप्रदजस्य गन्धकायितेच परस्परभेदज्ञा-
पनाय तत्तत्किट्टजन्यताबोधकसंज्ञया व्यवहार: क-
र्त्तव्य इति ॥

वक्ष्यमाणायां भागद्वयवत्यां सूच्यां वामभागे अ-
म्ला: किट्टानि तेषां संचित्तसंकेता: परमाणुगुरु-
त्वञ्च निहितं दच्चिणभागे फलभूतानि लवणानि

फलभूता: संच्छिप्तसंकेता: परमाणुगुरुत्वञ्च निहि-
तमस्ति यथा ॥

लघुतमाया जीवान्तकायितम्

जीवान्तकिकाम्ब्लं= जी+५प्रा=५४ । लघुतमा=ल+प्रा =४८ ।	लघुतमाया जीवान्तकायितम् =लप्रा+जी५प्रा=१०२ ॥

अयं लवणविशेषो ऽग्निशस्त्रप्रयोगोपयुक्तचूर्णनिर्मा-
णोपयोगित्वेन लोके प्रसिद्ध: ॥

अथ नवसागराख्यजनकस्य जीवान्तकायितम्

जीवान्तकिकाम्ब्लम्=५४ नवसागराख्यजनकम= ३ज+जी=१७	नवसागराख्यजनकस्य जीवान्तकायितम्=७१ ॥

अयं लवणविशेषो जीवान्तकस्य प्रथमप्राणप्रदजना-
मकं उन्मादकं पूर्वोक्तं वायुविशेषं जनयति ॥

अथ रजतस्य जीवान्तकायितम्

जीवान्तकिकाम्लम्=५४ ⎫ रजतस्य जीवान्तकायि-
रजतस्य प्राणप्रदजम्= ⎬ तम्=१७२ ॥
११०+८=११८ ⎭

अयं लवणविशेषश्चिकित्सकैः व्रणोत्पन्नदुर्मांसवि-
नाशाय प्रयुज्यते ॥

अथ लवणाकराया गन्धकायितम्

गान्धकिकाम्लम्=ग+ ⎫ लवणाकराया गन्धकायितम्
३प्रा=४० । ⎬ =७१ ॥
लवणाकरा=लव+ प्रा ⎪
=३१ ⎭

अयं लवणविशेषो विद्युद्यन्त्रनिर्माणे उपयुज्यते ॥

अथ चूर्णस्य गन्धकायितम्

गान्धकिकाम्लम् ⎫ चूर्णस्य गन्धकायितम्
=४० । ⎬ =६८ ॥
चूर्णम्=श+प्रा ⎪
=२८ ⎭

अयं लवणविशेषः कटिनीवत् शुक्लवर्णो भवति प्रतिमाविशेषनिर्माणे उपयुज्यते ॥

अथ ताम्रस्य गन्धकायितम्

गान्धकिकाम्लम्	ताम्रस्य गन्धकायितम्
$=४०$ ।	$=८०$ ॥
ताम्रस्य प्राणप्रदजम्	
$=$ ता$+$ प्रा$=४०$	

अयं नीलवर्णो लवणविशेषो विद्युद्यन्त्रनिर्माणे उपयुज्यते ॥

अथ लोहप्रथमप्राणप्रदजस्य गन्धकायितम्

गान्धकिकाम्लम्	लोहप्रथमप्राणप्रदजस्य
$=४०$ ।	गन्धकायितम्$=७६$ ॥
लोहप्रथप्राणप्रद-	
जम्$=$लो$+$प्रा	
$=३६$	

अयं हरिद्वर्णो लवणविशेषः लोके कासीस इति प्रसिद्धः ॥

अथ नवसागराख्यजनकस्य जलकरहरितायितम्

जलकरहरिताम्लम् =३६ । नवसागराख्यजनकः =१७ ॥	नवसागराख्यजनकस्य जलकरहरितायितम् =५३ ॥

अयं लवणविशेषो नवसागर इति प्रसिद्धो जलभ्री-
तीकरणादावुपयुक्तः ॥

अथ चूर्णस्याङ्गारिकायितम्

आङ्गारिकाम्लम् =अं+२प्रा= २२ । चूर्णम्=२८ ॥	चूर्णस्याङ्गारिकायितम् =५० ॥

अयं लवणविशेषो लोके कटिनीति प्रसिद्धः ॥

अथ लवणाकरायाष्टङ्कोपादानायितम्

टङ्कोपादानिका-
म्लम्=ट+६ प्रा
=६८ ।
लवणकरा=३१

लवणकरायाष्टङ्कोपादा-
नायितम्=८९ ॥

अयं लवणविशेषष्टङ्क इति प्रसिद्ध: ॥

अथ चूर्णस्य प्रकाशदायितम्

प्राकाशदर्किका-
म्लम्=प्र+५ प्रा
=७२ ।
चूर्णम्=२८

चूर्णस्य प्रकाशदा-
यितम्=१०० ॥

अयं लवणविशेषो मुख्यो मनुष्याद्यस्यारम्भको-ऽस्ति ॥

एवमुक्तप्रकारेण पूर्वोक्ततच्चानामप्रसिद्धान्यापामर-
प्रसिद्धानिच मिश्रणफलानि निरूपितानि तच्चानेक-
द्रव्ययोगादेकद्रव्योत्पत्तिर्वर्षिता । तदर्थं क्रियमाणोऽने-
कद्रव्यसंयोग: समासकरणमित्युच्यते । यच्च कस्यचिद-
वयविन आरम्भकद्रव्यजातिनिर्णयाय तदीयावयवानां

ज

विभाग: क्रियते तच्च स विभाग: समासकरणविपरीतत्वा-
द्व्यासकरणमित्युच्यते । तस्योपायं निरूपयितुमाह ॥

रसायनशास्त्रोक्तानां मिश्रणफलानामवय-
वविभागे त्रयो हेतव: औष्ण्यं विद्युत् द्या-
स्यद्रव्यारम्भकावयवगतपरस्पराकर्षणशक्त्य-
धिकत्वदन्यतराकर्षणशक्तिमत्पदार्थसान्नि-
ध्यञ्चेति ॥ २२ ॥

तच्चौष्ण्यं यथा पूर्वोक्तं रक्तवर्णं पारदस्य प्राणप्रदजं
अग्नितापेन विलीय शुद्धपारदरूपेण शुद्धप्राणप्रदवायु-
रूपेणाच व्यस्तं भवति ॥

प्राणप्रदवायो: शुद्धस्वरूपप्राप्त्यर्थमयमेक उपायो-
ऽस्ति यथाचैतत्तथैतदन्यव्याख्याने स्फुटीभविष्यति ॥

विद्युद्यथा विद्युत्संयोगेन जलं शुद्धजलकरवायुरूपेण
शुद्धप्राणप्रदवायुरूपेण च व्यस्तं भवति । विद्युत्साधन-
कव्यासकरणरीतिरेतत्प्रकरणटीकायां प्रदर्शयिष्यते ॥

व्यासद्रव्येत्यादि । यथा । पूर्वोक्तं तुल्याख्यं ताम्रस्य
गन्धकायितं स्वमिश्रितजले स्वच्छलोहखण्डप्रचेपेण
व्यसितुं शक्नुते । तथाहि । उक्तावस्थायां ताम्रम-
पेक्ष्य लोहस्याम्लाकर्षणशक्त्याधिक्यादम्लं लोहेन संयुज्य-
ते ताम्रञ्चाम्लेन वियुक्तं सत् लोहपृष्ठमेवाधिरोहति ।
सकले ताम्रे लोहपचाधिरूढे लोहांशेषुचाम्लमिश्रि-
तेषु जातेषु तच जले काशीशाख्यं लोहस्य गन्धका-
यितमुपलभ्यते नतु ताम्रस्येति ॥

रसायनज्ञैः प्रतिपदार्थं यावन्त आकर्षकाः पदार्थाः
प्रत्यचपरीचणैर्ज्ञायन्ते त एकस्मिंश्चक्रे लिख्यन्ते । तेषु
यो ऽधिकाकर्षकः स आदौ ततो न्यूनाकर्षक इति क्रमो
भवति । एतच्चक्रस्यावलोकनेन यस्मिन् कस्मिंश्चि-
त्पदार्थे तद्व्यासकरणार्थं ये ये पदार्थाः प्रचेप्यास्ते
ज्ञायन्ते ॥

एकोनविंशसूचव्याख्याने लिखितायां सूच्यां प्रदर्शिता
द्रव्याणां परमाणुगुरुत्वसंख्या तत्तुल्यबलाङ्कशब्दे-
नापि व्यवह्रियते यतस्ते मिश्रपदार्थोत्पादने स्वस्वाङ्क-

परिमितास्तुल्यबला भवन्ति । यच्च यः पदार्थः स्वाङ्क-
परिमितो वर्त्तते तचान्यो ऽपि स्वाङ्कपरिमाणेनैव स्थातुं
शक्नोतीत्यर्थः । यथा । पूर्वोदाहृतौ तुल्यस्य व्यासक-
रणेन चेदेकचिंशन्माषमितं ताम्रं पृथग्भवेत्तदा तच्च
सप्तविंशतिमाषमितं लोहं संयुज्येत यावत्त्वम्ब्ररसे एक-
चिंशन्माषमितं ताम्रं लीयते तावति सप्तविंशतिमाष-
मितमेव लोहं विलीयते नत्वधिकमित्यर्थः । एवं प-
रीचयेन ताम्रलोहयोस्तुल्यबले क्रमेणैकचिंशत्सप्तविं-
शत्यंशप्रमाणे भवतः । इत्थं प्रत्यच्चपरीचयैव चक्रेऽङ्का
लिखिताः । यस्य कस्यापि पदार्थस्य व्यासकरणार्थं नि-
र्मित्यर्थं वा तद्घटनावयवा यावत्प्रमाणा उपयुज्यन्ते
तत्प्रमाणानि चक्राङ्कालोकनेन निःशङ्कं चायन्ते रसा-
यनज्ञैः ॥

ताम्रगन्धकायितस्य लोहसंयोगाद् यो विकारो
व्यासकरणफलभूतः पूर्वमुक्तः स सङ्क्षिप्तसङ्केतोपायेन
प्रकाश्यते । यथा ॥

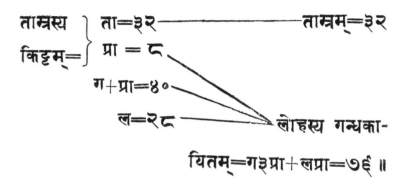

लवणजातिषु सर्वे धातवः किट्टरूपेणैव तिष्ठन्तीत्युक्तं ।
इदंत्ववधातव्यम् यद्रसायनीयपदार्थानां दर्शिततुल्यब-
लाङ्कानुसारेण यावानेव ताम्रसम्बन्धी प्राणप्रदो द्वाचि-
श्चत्वारिंश्मितं ताम्रं ताम्रगन्धकायिते किट्टतां नयति ता-
वानेवाष्टाविंशतिमितं लोहं किट्टतां प्रापयति परन्तु
जलमिश्रितगान्धकिकाम्ले प्रक्षिप्तं लोहं तदम्लेन स्वस्य
रसायनीयसंयोगसिद्धये यथोपयोगं जलात्प्राणप्रदमा-
कृष्य गृह्णाति । अत एकाकी जलकरो बुद्बुद्द्वारा
निर्याति इदञ्चानन्तरालेख्ये स्पष्टं यथा ॥

जलम् { जलकरवायुः —————————— जलकरवायुः
        प्राणप्रदवायुः
गान्धकिकाम्लम्
लोहम् ————————————→ लोहस्य गन्धका-
                            यितम् ॥

यैरुपायैरेताट्टश्बुद्बुदरूपः गुड्जलकरः रचितुं
शक्यस्ते उपायाष्टीकायां व्यक्तीकरिष्यन्ते ॥

द्रव्यचयविषयेषूक्तोदाहरणेषु सर्वत्रैकं द्रव्यमपरद्रव्य-
माकृष्य तृतीयमुत्सारयतीति समानं क्वचित्तु भिन्नभि-
न्नद्रव्यचतुष्टयारब्धे द्वे मिश्रणफले परस्परसंयोगदशा-
यां स्वघटकावयवादानप्रतिदाने कुरुतः । यथा चू-
र्णस्य जलकरहरितायितं यदा जलमिश्रितं सत्
तथाविधेन नवसागराख्यजनकस्य गन्धकायितेन संयु-
ज्यते तदा गान्धिकिकाम्लं चूर्णेन संयुज्यते जलकर-
हरितिकाम्लं च गान्धकिकाम्लवियुक्तेन नवसागराख्य-
जनकेन संयुज्यते । यदाच चूर्णस्य जलकरहरि-
तायितं केवलेन गान्धकिकाम्लेन संयुज्यते तदा तच्च

खसंयोगयोग्यद्रव्यान्तरानुपलम्भादायुरूपेण बहिर्निर्-
याति । तदिदमालेख्येन प्रदर्श्यते । तच सव्यभागमारभ्य
दचिणभागस्पर्शिनीभ्यां रेखाभ्यां एक: संयोग: प्रद-
र्श्यते सव्याद्दचिणभागमागच्छता बिन्दुसन्तानेनचापर:
संयोग: प्रदर्श्येत इति ।   तद्यथा ॥

नवसागराख्यज- ⎤	३ज+जी··········	·······नवसागराख्यजन-
नकस्य गन्धका- ⎬	ग+३प्रा	कस्य
यितम् ⎦		जलकरहरितायितम्
चूर्णस्य जलक- ⎤	ग्र+प्रा	चूर्णस्य
रहरितायितम् ⎦	ह+ज	गन्धकायितम् ॥

एतावता प्रकरणेन व्यासकरणोत्तरं मानुषशक्ति-
साध्यपुनरुत्पत्तय: पदार्था वर्णिता: । यथा । पारदस्य
प्राणप्रदज: प्राणप्रदरूपेण पारदरूपेणच यथा व्यस्त:
कर्तुं शक्य: तथा पुनरपि प्राणप्रदपारदयोर्मेलनात्पूर्व-
वत् कर्तुं शक्य इति ।   केचित्पुन: पदार्था नैताद्दश्व-
भावा: । यथा । काष्ठखण्डं व्यासकरणोपायै: प्राण-
प्रदजलकरशुद्धाङ्गारस्वरूपतां नेतुं शक्यते मांसखण्डं
वा प्राणप्रदजलकरजीवान्तकशुद्धाङ्गारस्वरूपतां न पुन:

६४

कयाचिदख्यत्युक्त्या पूर्वरूपमिति यत एषां पुनरूत्पा-
दने जीवनशक्तिरसाधारणं कारणं नच सा मानुषश-
क्तिसाध्येति ॥

जीवनशक्तीरसायनविधीन् प्रति-
बध्नात्यन्यथा वा करोति ॥ २३ ॥

भूमौ निक्षिप्तं बीजं मनुष्यशक्त्यसाध्यान् गुणान् वृक्षे
जनयति । यथा । मञ्जिष्ठाबीजं लोकोत्तरं वर्णमा-
स्वबीजं वा लोकोत्तरं रसमुत्पादयति ॥

तदेतस्माङ्केतोर्वनस्पतिविद्याविषयभूतान्वनस्पतिभे-
दान्निरूपयितुमारभते ॥

अथ वृक्षभेदज्ञापनाय सूचम् ॥

वानस्पत्या वनस्पतयश्चेति द्वि-
विधा वृक्षा भवन्ति ॥ २४ ॥

तच वनस्पतयो ऽपुष्पवन्त उच्छिलीन्ध्रादयः । पुष्प-
वन्तो वानस्पत्या आम्रादयः ॥

पिप्पलादयो दृढा अपुष्पवन्नरो न गण्यन्ते यतस्तेषां पुष्पाणि फलान्तर्भागे वर्त्तमानानि सूच्मदर्शकयन्त्रद्वारा दृश्यानि भवन्ति ॥

पुष्पाणां केसरा द्विविधा भवन्ति पौरुषा: स्त्रैणाश्चे- ति । येषामग्रेषु पुष्परजो वर्त्ते ते पौरुषा अन्ये स्त्रैणा भवन्ति ॥

वानस्पत्येषु समानपौरुषकेसराणां पृथग्वर्गा: क्रियन्ते तेषुच प्रत्येकं समानस्त्रैणकेसराणां वर्गा: पृथक्कल्प्यन्ते पचाद्याकारादिभिस्तेषां विशेषा निर्णीयन्ते ॥

अथ यदा पौरुषकेसररजांसि स्त्रैणकेसरे पतन्ति त- दा वीजमुत्पद्यते नल्वन्यथा । अतो यदि कस्यचित्पु- ष्पस्य पौरुषकेसरा निष्कास्येरंस्तदा तस्मात्फलं नोत्प- द्यते । यदि तु कस्यचिद्वृक्षस्य पुष्पात्पौरुषकेसरान- पास्य तत्स्त्रैणकेसरे तत्सजातीयान्यदृक्षपुष्पपौरुषकेसर- स्य रज: च्चिप्यते तदा तस्मात् तद्वृक्षद्वयगुणविशिष्टं वी-

जमुत्पद्यते । एवं यत्नेनानेके पुष्पविशेषा उत्पाद्यन्ते ॥

अथ कस्मिंश्चित्कस्मिंश्चित्सजातीयट्टच्चवर्गे कतिषु-
चिद्वृत्तेषु पौरुषकेसरविशिष्टान्येव पुष्पाणि भवन्ति ते
पुरुषाख्याः स्युः । अन्येषच ट्टच्चेषु स्त्रैणकेसरविशि-
ष्टान्येव पुष्पाणि भवन्ति ते स्त्रीसंज्ञकाः स्युः । तच्च
पुरुषाख्या नैव फलन्ति अत एव ते बन्ध्या उच्चन्ते ।
तेषां बन्ध्यत्वे स्त्रैणकेसराभावः कारणम्भवति । स्त्री-
संज्ञकाश्च पुरुषनिकटस्थाः पौरुषकेसररजःसंयोगेन
फलन्ति यथा खर्जूरादयः ॥

ट्टच्चविशेषा लच्चसंख्याधिका भवन्ति किमिद्द तेषां
वर्णनेनेत्यलम् ॥

अथ ट्टच्चावयवसंस्थितिवर्णनं शरीरवर्णनप्रकरणे क-
रिष्यते । ट्टच्चा जीवनविशिष्टा अपि स्थामान्तरगम-
नासमर्था इति ते जन्तुभ्यो भिद्यन्ते । अथ जन्तु-
वर्णनम् ॥

जन्तुभेदच्चापनाथ सूचम् ॥

जन्तूनां चत्वारो वर्गा भवन्ति पृष्ठ-
वंशविशिष्टानां कोमलशरीरविशि-
ष्टानां काण्डविशिष्टानां शरीराभि-
तोऽवयवविशिष्टानाच्चेति ॥ २५ ॥

तच प्रथमवर्गजाताः पृष्ठवंशविशिष्टा यथा मनु-
ष्यादयः । द्वितीयवर्गजाता अतिकोमलशरीरविशि-
ष्टाः प्रायः शुक्त्यन्तर्गता भवन्ति यथा शम्बूकादयः ।
तृतीयवर्गजाताः काण्डविशिष्टशरीरा यथा शतपदा-
दयः । चतुर्थवर्गजाता अभितोऽवयवविशिष्टा यथा
जलनीलीव्याप्तसलिले सञ्जाता ह्रस्वा प्रायः सूक्ष्मदर्श-
कयन्त्रेण दृष्या जन्तुविशेषाः ॥

अथ प्रथमवर्गस्य चत्वारो भेदा भवन्ति तच प्रथमे
सस्तना द्वितीये पक्षिणस्तृतीये सर्पाश्चतुर्थे मत्स्याः ॥

अथ सक्तनाः पुनर्द्वादशधा भवन्ति । ते यथा द्वि-
हस्ता मनुष्याश्चतुर्हस्ता वानरा हस्ताभ्यां उड्डीयमाना
वातुल्यः कीटभोजिनश्छुच्छुन्दर्यादयो मांसभोजिनो
व्याघ्रादयो जलस्थाः शिशुमारादयः स्थूलचर्म्मिणो ग-
जादयो रोमन्यकरा एषादयो अदन्ता वज्रकीटादयः
कर्त्तनकरा मूषिकादयः पेशीविशिष्टाः कङ्गरूसंज्ञका
जन्तुविशेषाश्चञ्चुविशिष्टाश्चतुष्पदा अस्त्रेलियाख्ये देशे
निवसन्ति ॥

अथावशिष्टानां चयाणां वर्गाणां प्रत्येकं बहवो भेदा
भवन्ति ते विस्तरभयादिह नोच्यन्ते ॥

अथ जन्तूनां एच्चागाश्च शरीरिल्वरूपो धर्म्मोऽन्यव-
स्तुभ्यो भेदक इति पूर्वं सूचितं । तच्च जन्तूनां शरीरवलं
प्रसिद्धं एच्चागां शरीरिल्वन्तु एङ्गादिरूपचेष्टाहेतुजला-
कर्षणकर्त्तृमूलादिकर्म्मेन्द्रियवच्चात् । तदुक्तं गौतमेन ।
चेष्टेन्द्रियार्थाश्रयः शरीरमिति । चेष्टाया इन्द्रियाणां

सुखदुःखादिरूपार्थीनां चाश्रय इत्यर्थः । परन्तु
इच्छायां ज्ञानेन्द्रियसच्चासच्चयोर्विनिगमनविरहात्सु-
खादिमत्त्वं न निश्चितम् । तच्च पूर्वप्रतिज्ञातं इच्छादी-
जोत्पत्तिप्रकारमभिधातुमाह ॥

भूमिनिहितं बीजं स्वखजातीयाङ्कु-
रोत्पादानुकूले काले ऊर्ध्वाधोगा-
मिनौ द्वावङ्कुरौ जनयति तच्चोर्ध्व-
गामिना प्रकाण्डो ऽधोगामिनाच मू-
लमारभ्यते ॥ २६ ॥

भूमध्यान्निर्गच्छन्नवाङ्कुरस्तेजसे स्पृहयतीव यतस्ते-
जःसंयोगाभावे हरितवर्णरहितपच्चतां प्राप्ता इच्चजा-
तिर्म्लायते । ययैव दिशा यावतैवावकाशेन तेजस्तन्-
निधिं प्राप्नोति तामेव दिशं तावतैवावकाशेन इच्चजा-
तिरपि तेजोऽभिमुखं याति ॥

इच्चाखशरीरपुष्टिहेतून् पदार्थान्

मूलेन पृथ्वीतः पञ्चैश्च वायोः सका-
शादाकृष्य गृह्णन्ति ॥ २७ ॥

प्राणप्रदजलकराङ्गारा ह्यचजातेस्साधारणारम्भका
इति पूर्वं सूचितम् । तच ह्यचहट्द्विविषये जलसंकस्यात्या-
वश्यकत्वेन ह्यचैस्स्वमूलाकृष्टजलात्प्राणप्रदजलकरौ प्रा-
प्येते । अङ्गारस्तु वायोः सकाशात्प्राप्यते यतोऽङ्गारः
प्राणप्रदमिलितोऽङ्गारिकाम्लवायुविशेषस्वरूपेण साधा-
रणवायौ तिष्ठति सच वायुविशेष उच्छासक्रियया का-
छादिदहनक्रियया चोत्पद्यते । प्रत्यचन्वाच प्रमाणम् ।
तथाहि । काचपाचान्तर्वर्त्तिनि स्वच्छजलसटुग्मे चूर्ण-
जले काचनलिकाद्वारा श्वासवायुः प्रवेश्यतां येन ख्वा-
सक्रियाजन्याङ्गारिकाम्लवायुसंयोगवशात्तज्जलान्तर्गत-
चूर्णे कठिन्यवस्थामागच्छति सति तज्जलं कठिनी-
मयत्वात् चीरवर्णं भविष्यतीति ॥

जन्तूनामुदरफुफ्फुसे यत्कार्यं सम्पाद-

यतः तदेव कार्यं वृक्षाणां पर्णानि
सम्पादयन्ति ॥ २८ ॥

यथा जन्तूनामुदरं भुक्तान्नसम्बन्धिनां शरीररक्षा-
सम्पादकानामंशानां अन्तःकुङ्क्षेणानुपयुक्तानामंशानां
बहिर्निःसारणेनचोपकारकं भवति तथैव वृक्षाणां पचा-
र्युक्तकार्यसम्पादनेनोपकुर्वन्ति । वृक्षाणां स्तम्बशा-
खाद्युत्पादनायाङ्गार आवश्यकः सच पर्णैराङ्गारिकाम्ल-
वायुसकाशादाह्रियते तत्सम्बन्धिप्राणप्रदश्च त्यज्यते ॥

तदेवं जन्तवस्खसजातीयापकारकं वृक्षाणामत्युप-
कारकं वायुं निःश्वसन्ति वृक्षा अपि जन्तुनिःश्वसितं
स्वोपकारकं वायुमन्तरादाय जन्तूपकारकं प्राणप्रदवा-
युं बहिर्निःसारयन्तीतिसिद्धम् ॥

आङ्गारिकाम्लवायुर्जन्तुभिर्निःश्वस्यत इत्यच पूर्वं युक्ति-
र्दर्शिता इदानीं प्राणप्रदो वृक्षैर्निःश्वस्यत इत्यच प्रमाणम्

प्रदर्घ्यते । तथाहि । अम्लानपञ्चसहितं जलपूर्णं विशा-
लकाचपात्रं जलपूर्णं पात्रान्तरेऽधोमुखं संस्थाप्यातपे
स्थाप्यतां ततः पत्रेभ्यः प्राणप्रदवायुर्निर्गमिष्यति तन्नि-
र्गमनलच्छणानिच पूर्वोक्तरसायनप्रकरणप्रतिपादितवि-
ध्यनुसारेणावगन्तव्यानीति ॥

द्टचाणां पुष्यनिर्मितिव्यवस्था तदीयसकलावयव-
प्रयोजनञ्चेत्येतद्द्वयं स्त्रीपुञ्जातीयकेसरानुसारेण द्टच्च-
भेदनिरूपणावसरे संच्छेपेण वर्णितम् ॥

अथेदानीं जन्तुशरीरसंस्थितिर्विविच्यते तचादौ म-
नुष्यशरीरावयवव्यवस्थायां सूचम् ॥

अस्थिपञ्जरो मांसरज्जुर्मस्तिष्कमुदरं
हृदयः फुप्फुसं रक्तवहा नाड्य इत्य-
वयवा मनुष्यशरीरस्य ॥ २८ ॥

तत्र चतुःपञ्चाशदधिकद्विशतसङ्ख्याकास्थिसङ्घातो ऽस्थिपञ्जरः इदञ्चैतत्प्रकरणव्याख्यायां स्फुटीभवि- ष्यति ॥

मांसस्य मुख्योऽशो मांसरज्जुः । अस्या आकुञ्चनप्र- सारणे इच्छाजनितयत्नेन भवतः ताभ्याञ्चास्थिपञ्जरव- र्तीन्यस्थीनि चाल्यन्ते ॥

विशेषतः शिरसि वर्तमानः शुक्लवर्णः स्नेहविशे- षो मस्तिष्कं । सूक्ष्मतन्त्वाकारा अस्यांशाः पृष्ठवंशद्वा- रा प्रसरन्तो नखकेशान्विहाय सर्वान्शरीरावयवान् व्याप्नुवन्ति । अत एव नखकेशावच्छेदेनैच्छिकी क्रिया ज्ञानञ्च नोत्पद्यते ॥

शरीरपुष्ट्युपयुक्तानुपयुक्तावन्नांशौ येन विभाज्येते तदुदरम् ॥

जीर्णोमन्नं सर्वावयवपुष्ट्युत्पादकरुधिरवर्णकशुक्लर- सस्वरूपेण परिणतं नालद्वारा यत्र प्रविशति स हृदयः ॥

अ

सन्ततांकुञ्चनप्रसारणवतो हृदयस्थलान्निर्गत्य यच्च रुधिरं प्रथमं प्राप्नोति तत्फुप्फुसम् । हृदयनिर्गतं रुधिरं यद्द्वारा फुप्फुसं प्रविशति पुनश्च सर्वशरीरे कृतसञ्चारं पुनर्यद्द्वारा हृदयं प्राप्नोति ता नाड्यो रक्तवहा इत्युच्यन्ते ॥

फुप्फुसदेशे रुधिरं प्राणप्रदवायुना संयुज्यते अङ्गारेण वियुज्यतेच एतद्विपरीतस्वभावा त्वचजातिरङ्गारेण संयुज्यते प्राणप्रदेन वियुज्यतेचेति पूर्वं सूचितम ॥

एतावता भूगोलघटकावयवा यथात्मानं निरूपिता: इदानीं तद्वयवानां प्राप्तिस्थानानि निरूप्यन्ते ॥

भूपृष्ठस्थनदीसमुद्रपर्वतादयो भूपृष्ठविद्याप्रसङ्गे संक्षेपेण वर्णिता: । तच्च सर्वपर्वतानामारम्भकद्रव्याणि नैकजातीयानि क्वचिद्धिमालयादौ पर्वतेऽतिकठिनानां क्वचिच्च सैकतपाषाणपर्वतादौ मृद्वीनां शिलानामुपलभ्मात् । हीरका: प्रस्तराङ्गरास्च सर्वभूमिषु साधारणेन

प्राप्तियोग्या न भवन्ति अपितु तत्प्राप्तिस्थानानि विशेष-
लच्चणलच्चितानि सन्ति इत्येतत्सर्वं व्याख्याने स्पष्टीकरि-
ष्यते । पार्थिवविषयभूतानामाकरजपाषाणानामवय-
वव्यवस्थाज्ञानपूर्वकेणाभीप्सितपाषाणविशेषोपलम्भस्य-
लच्चानेन साध्यस्य तज्ज्ञातिप्रत्यभिज्ञानस्य सिद्ध्यर्थं पाषा-
णविद्याया उपक्रमादादौ पाषाणानां भदेप्रदर्शनार्थं
सूत्रम् ॥

नियताकारा अनियताकाराश्चेति पा-
र्थिवांश द्विविधा भवन्ति ॥ ३० ॥

नियताकारा हीरकपद्मरागमरकतादयः । अनि-
यताकाराः कर्करमृद्दादयः ॥

नियताकारा बहुधाकारदृष्टिमात्रेण एतद्विद्याज्ञानि-
भिर्निश्चीयन्ते ॥

रूपगुरुत्वादिलिङ्गैरनियताकाराणाम्भेदा ज्ञायन्ते र-
सायनशास्त्रव्यासकर्तृणोनच ॥

यथा पलाण्डुरनेकैस्त्ववपुटैराटतो
भवति तद्वद्भूरप्यनेकैः पार्थिवांश-
पुटैराटता भवति परं तानि पार्थि-
वांश्पुटानि प्रायश्भिच्छन्नभिन्नानि
जातानि ॥ ३१ ॥

यथा । पृथिव्यां क्वचित्प्रदेशे कठिनीमयानि भूपुटानि
जलाश्रयतलस्थितपङ्कवत्समपृष्ठानि भवन्ति प्रदेशान्तरे
पुनरधरपुटनिर्गच्छच्छिलासमूहभग्नावस्थानतया विष-
मपृष्ठानि भवन्ति । तत्र यो हेतुः सोऽग्रिमप्रकरणे वि-
चारयिष्यते ॥

अस्त्येतस्या विद्याया लोके महानुपयोगः । तथा-
हि । एतद्विद्याया ज्ञानिनः प्रत्येकं पार्थिवांश्पुटानां
निर्मितिं तत्तद्वत्यंस्थां इच्चांशानाञ्च स्थितिं निरीक्ष्य
पार्थिवांश्पुटस्थितिक्रमं निरनयन् अत एतद्विद्याविद्
यत्पार्थिवांश्पुटं खनिजाङ्गारपार्थिवांश्पटात्सदा ऽधरेव

वर्तते तच्च खनिजाङ्गारलाभाय खनन्तं मनुजं वीक्ष्य इह खनिजाङ्गारं मा खगयस्व तवेह श्रमो वृथा स्यादिति तं कथयित्वा तस्य निष्फलदुःखं दूरीकुर्यात् पुनर्यदि ताट्टशं पार्थिवांश्रुपुटं पश्येदधस्तात्खनिजाङ्गारो व-र्तते तर्हि तच्च खननार्थं खनिजाङ्गारान्वेषणमचि-कीर्षन्तं तं मनुजं बोधयेत् । एवं ये ये उपयोगिनो भूम्यन्तर्गताः पदार्था यच्च यच्च वर्तेरन् ते ते पदार्था-स्तत्तत्स्थानञ्वैतद्विद्याविद् जानाति ॥

आसां सकलप्रतिज्ञानां प्रमाणानि अस्मिन् संक्षित्र-ग्रन्थे वक्तुं न शक्यन्ते । तथाप्यासां मण्डनं खण्डनं वा बुद्धिमद्भायोग्यं न स्यात् ॥

पृथिव्यां कतिपयेषु स्थानेषु यच्चाधुना स्थलमस्ति तच्च पूर्वं समुद्र आसीत् । यच्च समुद्रस्तच्च कदाचित् स्थ-लमासीत् । तदानीं ये प्राणिनो वृच्चाश्वाच न्यवसन् ताट्टशा अधुना न दृश्यन्ते ॥

प्राचीनजन्तूनां विशेषा आश्चर्यकरा भवन्ति । ये

ज्ञानिनो यंकञ्चन प्राणिनं तदेकास्थिनिरीचणमाचेण
कदाचित् परिचिन्वन्ति किम्पुनस्ते प्राक्तनजन्तून् तद्-
स्थिपञ्जरावलोकनेन जानन्तीति वाचम् । अथ पा-
षाणान्तर्गतप्राणिनां वर्णनमतिसंचेपतः क्रियते ॥

प्राचीनकालिको हस्ती हस्तषट्काधिकोच्छायो ह-
स्तैकादशकाधिकदैर्घ्यविशिष्टो रोमाक्रान्तशरीरश्वासी-
त् । तस्य ट्टहद्न्तावतिकुटिलौ दैर्घ्ये करषट्कमिता-
वास्ताम् ॥

प्राचीनकालिको महामकरो दैर्घ्ये चयस्त्रिंशत्कर आ-
सीत् । तस्यास्थीनि इङ्गलण्डदेशे प्रामुवन्ति यच्चाधुना
शीतवशान्न कश्चन मकरः स्थातुं शक्नोति ॥

प्राक्तनी महागोधा सप्तचत्वारिंशत्करदैर्घ्या ऽभ-
वत् ॥

अथ मत्स्यमकरस्वरूणाभावेन पच्चाभ्याञ्च मत्स्यसदृशे
विंशतिकरदैर्घ्यश्वाभवत् । नरशिरःपरिमिते तस्याचि-
णी आस्ताम् ॥

अथ मकरकल्पसंज्ञको मत्स्यमकरेण किंचित् सट्ट-
शस्तस्य शिरो ऽत्यन्तमल्पं ग्रीवाच सर्पवद्दीर्घा ऽभवत् ॥

उड्डीयमानमकरस्य मुखं मकरमुखवत् पच्छद्वयं वा-
तुलिपच्छवदभवत् ॥

एतावता पृथिवीस्थपदार्थानां या व्यवस्था दर्शिता
तस्याः कारणानि निरूपयितुं पार्थिवांश्यव्यवस्थाहेतु-
विचाराख्यं प्रकरणमारभ्यते ॥

भूगोलमध्यभागो ऽग्निना द्रवीकृ-
तोऽस्तीति सम्भाव्यते ॥ ३२ ॥

इदञ्च वक्ष्यमाणानुमानेन कल्प्यते । यदि भूम्यु-
ष्णता सूर्यसम्बन्धादेव भवेत्तर्हि भूपृष्ठे तदाधिक्येन
भवितव्यं गम्भीरखन्यादौ तन्यूनताच स्यात् । परन्तु
वस्तुत इदं वैपरीत्येनानुभूयते गाम्भीर्याधिक्यानुसा-
रेण खन्यादरुष्णताधिक्यदर्शनादिति ॥

बहुषु स्थानेष्वाग्नेयाः पर्व्वता वर्त्तन्ते । तेषामुत्प-
त्तिरेवम्भवति पृथिवीमध्यादागतो द्रवीभूतः पार्थिवां—
श्च ऽत्यन्तवेगेन पूर्वस्थितम्पार्थिवांशपुटं स्फाटयति तिर्य-
क्करोतिच । ततस्तच्छेदाद्वहिर्निर्गत्य पर्वतो भवति ॥

तत्पर्वतशिखरे मेघाः समागच्छन्ति ततश्च सरित
उत्पद्यन्ते ॥

सरितः प्रवाहाघातैस्तत्पर्वतस्यावयवाश्चूर्णीभूय ज-
लमिश्रिताः समुद्रङ्गच्छन्ति । तत्र सम्मेदे सरितो
जलं स्थिरम्भवति तेन तन्मिश्राश्चूर्णीभूतपर्वतावयवा
अधः सञ्चीयन्ते ततस्तचैकमल्पद्वीपमुत्पद्यते नद्याश्च
मुखद्वयं जायते ॥

पुनः प्रतिमखमेवमेकैकमल्प द्वीपमुत्पद्यते एवमने—
कानि सरिन्मुखानि जायन्ते ॥

अनयैव युक्त्या गङ्गायाः शतमुखान्युत्पन्नानि ।
तन्मुखव्याप्तस्थानं सुन्दरवमाख्यम् ॥

तच जाते एकैकाल्पद्वीपे बहवो दृच्छा जन्तवस्तीनिच
समायान्ति प्रतिद्वीपतटं शङ्खविशेषाश्चायान्ति ॥

पुनर्यदि तच ज्वलनपर्वत उत्पद्यते तर्हि तच स देश
उच्छितो भवेत् शङ्खादीनि सामुद्रवस्तूनिच तच ल-
भ्यानि भवेयुर्यथा बहुषु पर्वतशिखरेषु लभ्यन्ते ॥

नैतत्सर्वमल्पकालेन सम्पद्यते अतो ऽल्पायुषा मनु-
जेन नेदन्तदारम्भकालाद्यावधि अवलोकयितुं शक्यते
किन्तु यथा विचिचपरिमाणान् वनदृच्छानवलोक्य ये
महान्तो दृच्चास्ते पूर्वमल्पा आसन् येचाल्पास्ते महान्तो
भविष्यन्तीत्यनुमीयते तद्वदल्पसरित्सङ्गमे समुत्पन्नस्या-
ल्पद्वीपस्यानुदिनं विकारं तच्चत्यविचिचपदार्थाश्च दृष्ट्वा
ऽन्यच वर्तमानस्यैताटृक्भूप्रदेशस्योत्पत्ते: प्रोक्तकारणम-
न्तरेणान्यत्कारणं न स्यादित्यनुमीयते ॥

सुन्दरबनभूमिर्यदि भूम्यन्तर्गतवह्निनोच्छिता भवेत्
तदा तद्देशपूर्ववर्तिनां व्याघ्रमृगमकरादीनामस्थीनि
पृथक् संयुक्तानिच तन्मृदि मग्नानि प्राप्नुयु: । तथा

प्राचीनकाले ये देशा एवमुत्पन्नास्तिष्वपि तत्प्राचीनका-
लिकानां जन्तूनामस्थ्यादीनि लभेरन् । वस्तुत एवं
लभन्ते यथा पूर्वेप्रकरणे उक्तम् । अथैतावता अस्म-
द्विचारविषयाणामेतेषां प्रस्तुतपदार्थानां तत्त्वनिर्ण-
यस्य प्रत्यक्षमाचेणासम्भवादिदानीमनुमानाख्यप्रमाणा-
न्तरमभिधातुमुपक्रमते ॥

❋ ॥ इति चतुर्थोध्यायः समाप्तः ॥ ❋

CORRIGENDA.

—o-o-o—

At Page 1 line 4 *For* " First Book" *read* Second Book.

„ „ 8 „ 30 Subjoin the words—" Leibnitz, who made the discovery at the same time, is called the Differential and Integral Calculus."

„ „ 9 „ 9 *For* " comparative" *read* respective.

„ „ 10 „ 10 *For* " line" *read* time.

„ „ 23 „ 15 *For* " simple" *read* simple multiple.

„ „ 37 „ 21 *For* " and ascertain Nitrogen, and Carbon; that it consists of Oxygen, Hydrogen," *read* and ascertain that it consists of Oxygen, Hydrogen, Carbon, and Nitrogen.

„ „ ५६ „ ८ *For* " अङ्गारिकायितन्" *read* अङ्गारायितम्,—and so elsewhere, wherever the इक of the name of the Acid has been inadvertently retained in the name of the Salt.

„ „ ८२ „ ७ *For* "चतुर्थाध्यायः" *read* तृतीयाध्यायः ।

Milton Keynes UK
Ingram Content Group UK Ltd.
UKHW032319161024
449665UK00001B/52